LIFE SCIENCES MISCELLANEOUS PUBLICATIONS
ROYAL ONTARIO MUSEUM

Rolf Ludvigsen

Fossils of Ontario
Part 1: The Trilobites

Publication date: 31 January 1979

ISBN 0-88854-221-6

ROYAL ONTARIO MUSEUM
PUBLICATIONS IN LIFE SCIENCES

The Royal Ontario Museum publishes three series in the Life Sciences:

LIFE SCIENCES CONTRIBUTIONS, a numbered series of original scientific publications including monographic works.

LIFE SCIENCES OCCASIONAL PAPERS, a numbered series of original scientific publications, primarily short and usually of taxonomic significance.

LIFE SCIENCES MISCELLANEOUS PUBLICATIONS, an unnumbered series of publications of varied subject matter and format.

All manuscripts considered for publication are subject to the scrutiny and editorial policies of the Life Sciences Editorial Board, and to review by persons outside the museum staff who are authorities in the particular field involved.

ROLF LUDVIGSEN is a Research Associate of the Department of Invertebrate Palaeontology, Royal Ontario Museum, and Assistant Professor in the Department of Geology, University of Toronto.

Cover drawing: *Arctinurus boltoni* (Bigsby)

100 Queen's Park, Toronto, Canada M5S 2C6
PRINTED AND BOUND IN CANADA BY THE HUNTER ROSE COMPANY

Contents

Foreword

More than 100 years have passed since H.A. Nicholson, Professor of Natural History in the University of Toronto, published in 1874 and 1875 his two reports on the palaeontology of the province of Ontario. Since that time not only has there been a great increase in the volume of palaeontological research on Ontario, but increasingly, publication has been in diverse and sometimes difficult to obtain journals. The effects of this have been to make it difficult for the professional, as well as for the amateur, to obtain up-to-date information.

This new series, *Fossils of Ontario*, seeks to remedy some of these problems by drawing together and synthesizing information on particular taxonomic groups as they occur in the rocks and sediments of Ontario. The authors of each contribution are experts in the particular taxonomic group or topics being considered and their synthesis is directed to the student, the collector, the professional palaeontologist, as well as to the interested person who wishes to be informed.

P.H. von Bitter,
Associate Curator-in-charge,
Department of Invertebrate Palaeontology,
Royal Ontario Museum.

Associate Professor,
Department of Geology,
University of Toronto.

Acknowledgements

I am grateful to the curators of the various institutions who responded promptly to my requests for particular trilobite specimens and especially to the staff of the Royal Ontario Museum and the Geological Survey of Canada who gave me free access to their trilobite collections. Richard Fortey of the British Museum (Natural History) supplied the photographs of *Isotelus platycephalus*, Murray Copeland of the GSC brought to my attention the cock-eyed specimen of *Phacops rana*, Allen Ormiston of Amoco Research helped in the identification of several Devonian trilobites, and Don Mikulic of Oregon State University did the same for Silurian trilobites.

David Dillon executed the fine trilobite reconstructions seen on the cover and in the text, and Subhash Shanbhag drafted the figures. I am especially indebted to Brian O'Donovan for his expert photographic work and to Desmond Collins and Peter von Bitter of the ROM for their encouragement during the preparation of this book.

The cost of photography was partially defrayed by the Department of Invertebrate Palaeontology of the Royal Ontario Museum.

Finally, I thank Tom Bolton and Niles Eldredge for reading the manuscript.

Introduction

Within the layered Palaeozoic rocks of Ontario are preserved the remains of an extinct group of animals whose demise predated the appearance of the human animal by over 200 million years and whose name, along with that of the dinosaurs, has become synonymous with the term *fossil*. These creatures are trilobites, and their fossil record in Ontario is the subject of this book.

Trilobite remains are often fascinating and aesthetically pleasing. Their peculiar shapes and striking ornamentation attract the interest of those who collect the strange, the unusual, and the beautiful; the minute perfection of their petrified outer skeletons delights those who seek natural beauty. However, trilobites are not merely interesting objects; they are also valuable in palaeobiologic contexts in that they constitute a critical record of a unique group of well-adapted and highly successful marine arthropods. This record has yielded a surprising amount of anatomical, evolutionary, ecological, zoogeographical, and even behavioural information about these extinct animals.

Trilobites are among the most lifelike of fossils, and perhaps this is one explanation of their popular appeal. Little imagination is required to "revive" a trilobite. Many well-preserved, complete specimens appear to belie their great antiquity; some almost seem ready to arch their bodies, peer about, and crawl forward as if to continue a journey that was interrupted in the Palaeozoic.

The first step towards the appreciation and understanding of any natural object is identification. Recognition of a peculiar blob in a rock as a fossil offers limited information about its nature and significance. If, however, the identification of this blob can be extended from fossil to trilobite, to lichid trilobite, to *Arctinurus*, and finally to *A. boltoni* (Bigsby), then each step in identification increases the cumulative awareness of the significance of this relic of past life. Thus the blob becomes an item of scientific importance through the process of identification.

The Palaeozoic rocks of Ontario contain more than 60 genera of trilobites and perhaps a few hundred species. These trilobites have been described and illustrated in various professional journals, government publications, and scholarly monographs, but most of these are difficult to obtain and a few are more than a century old. The palaeontologic information included in these works is generally unavailable to the serious amateur collector, the natural historian, and the student of geology and palaeontology.

This publication provides a reference for identifying Ontario trilobites through good illustrations and brief descriptions of each genus. I hope that the illustrations will whet the reader's interest to the extent that he or she will grab hammer and hand-lens and go forth to the outcrops, roadcuts, and quarries of

Ontario in search of trilobites. To aid these expeditions, a brief introduction to the Palaeozoic geology of Ontario and a series of maps showing the distribution of Palaeozoic formations is included. Because some readers, particularly students of geology and palaeontology, will want to search out additional information about certain trilobites, I have included a fairly extensive list of references. Although the book has been written with the collector and the student in mind, it may also assist palaeontologists from other parts of Canada and from other countries who need a quick overview of the Ontario trilobite faunas.

Naturally, I have selected the best-preserved and most complete of the Ontario trilobites for illustration, but when better specimens have been available from adjoining provinces and states, I have let the quality of the specimen take precedence over political boundaries. As a result, some of the "Ontario" trilobites in the illustrations were actually collected as far afield as Kentucky and Manitoba.

Exceptionally preserved specimens of trilobites are becoming increasingly difficult to find in Ontario, but while one is waiting for that perfect trilobite to come into view, there are thousands of more ordinary but still interesting specimens of most species to be collected and studied.

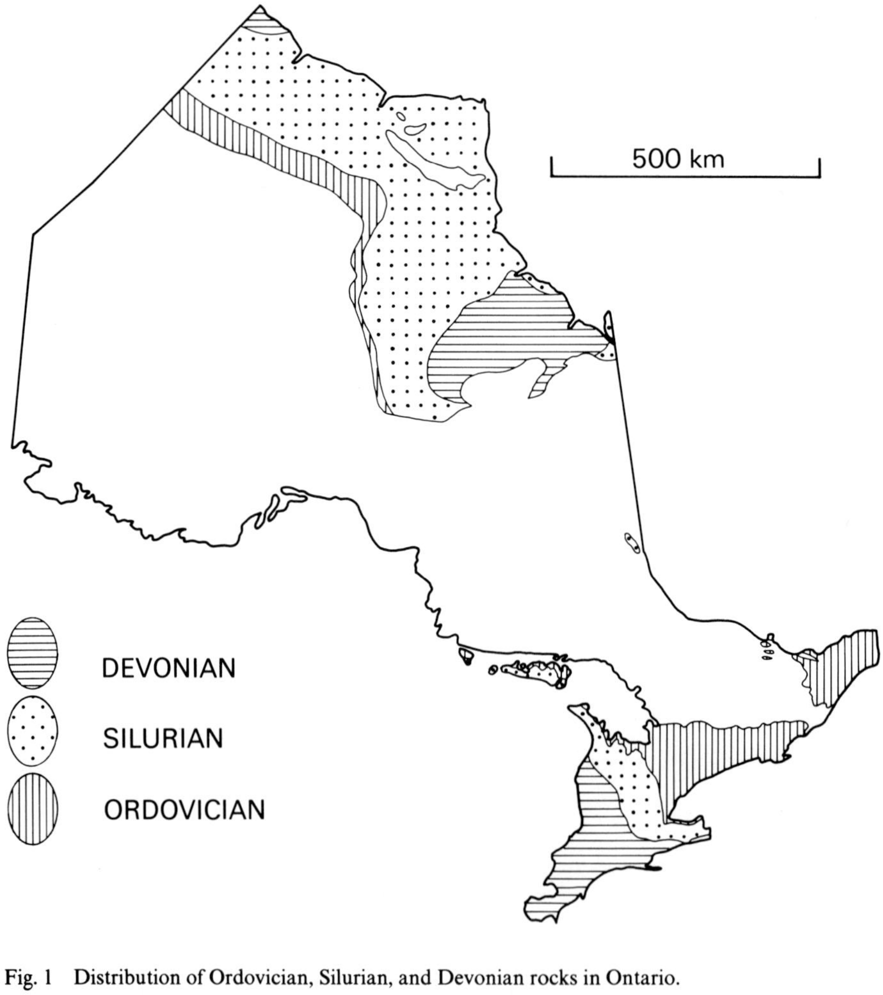

Fig. 1 Distribution of Ordovician, Silurian, and Devonian rocks in Ontario.

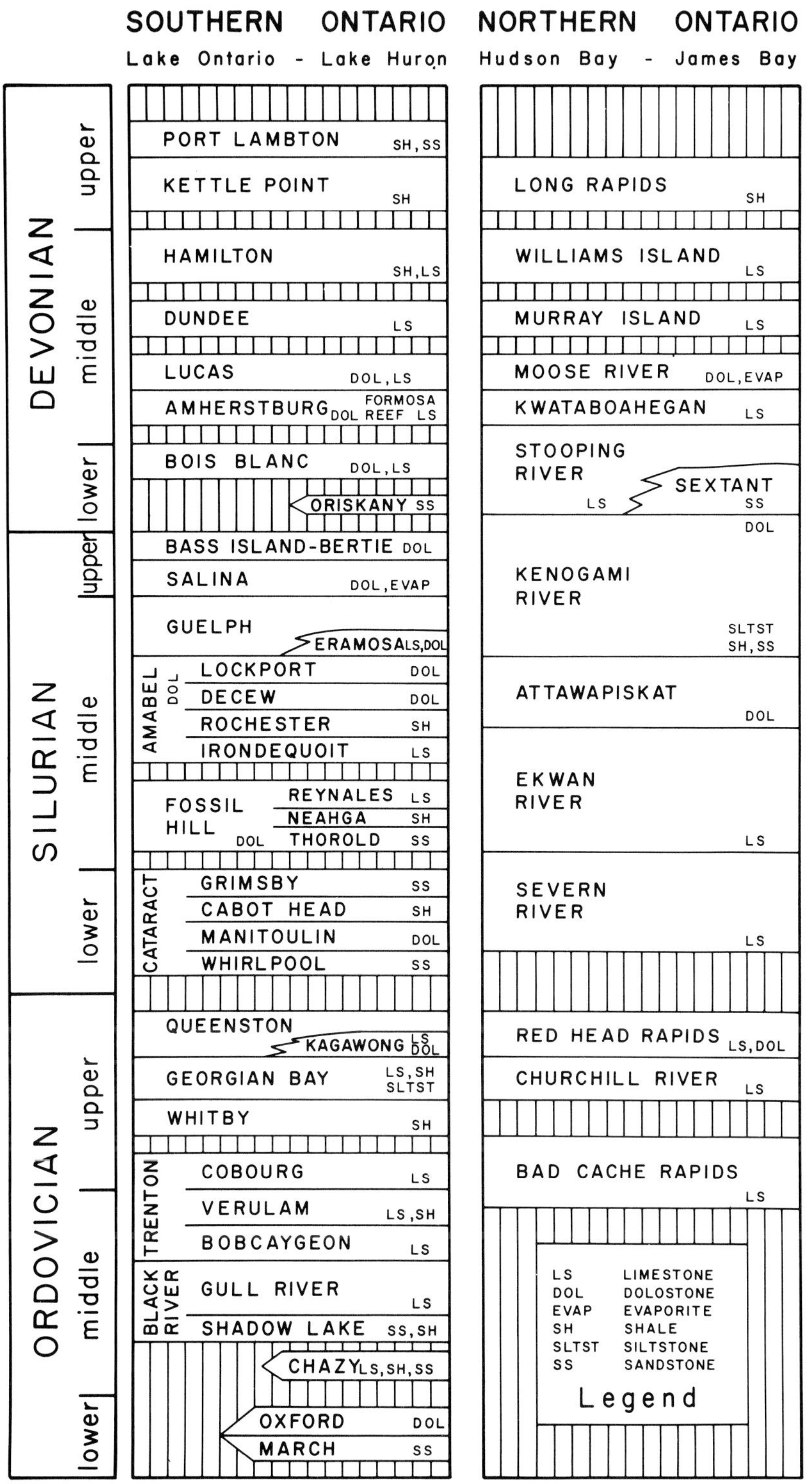

Fig. 2 Succession of Palaeozoic formations in southern and northern Ontario. Adapted from Winder and Sanford (1972) and Sanford and Norris (1973).

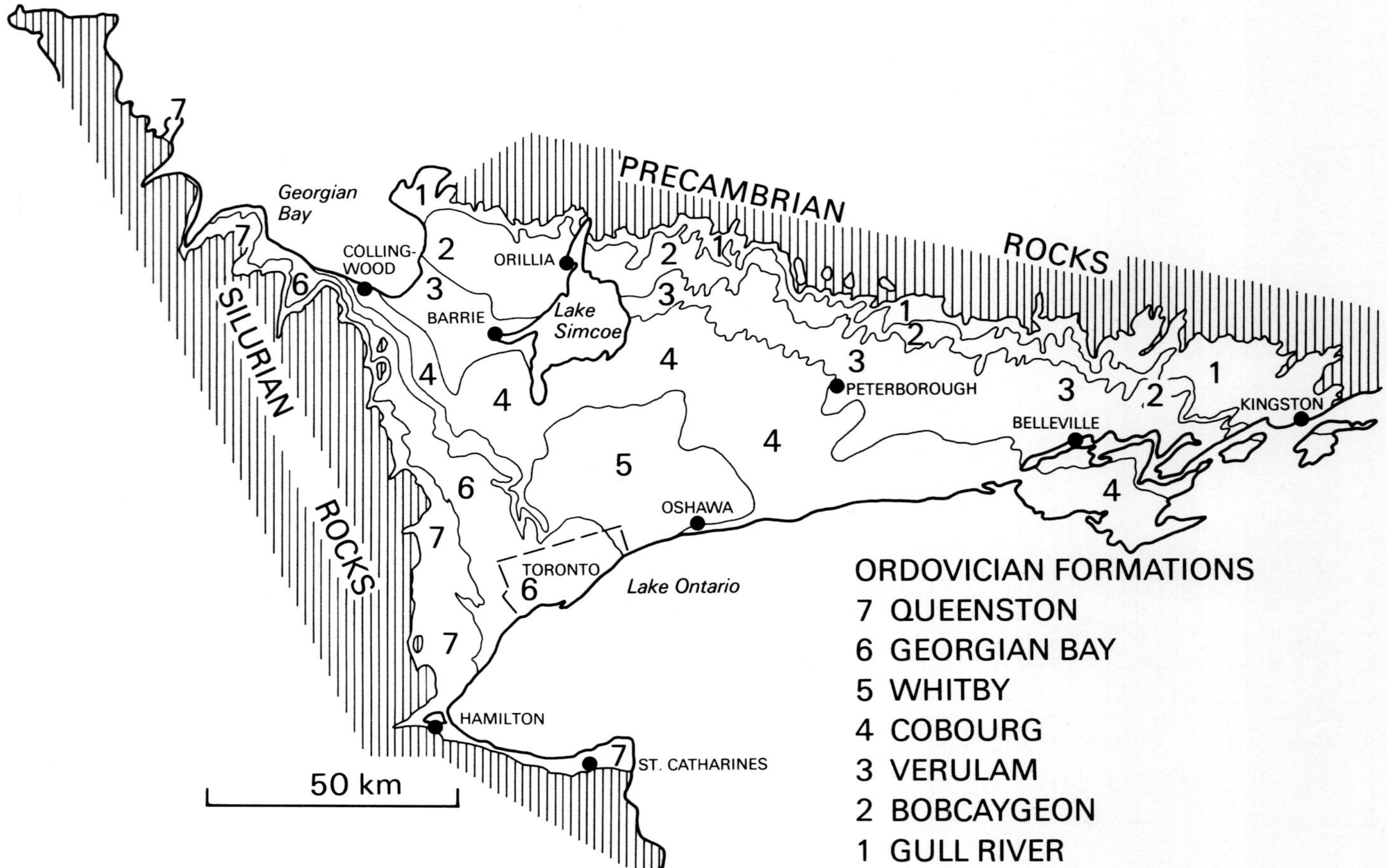

Fig. 3 Ordovician geology of southern Ontario. Adapted from Winder and Sanford (1972) and Liberty (1964).

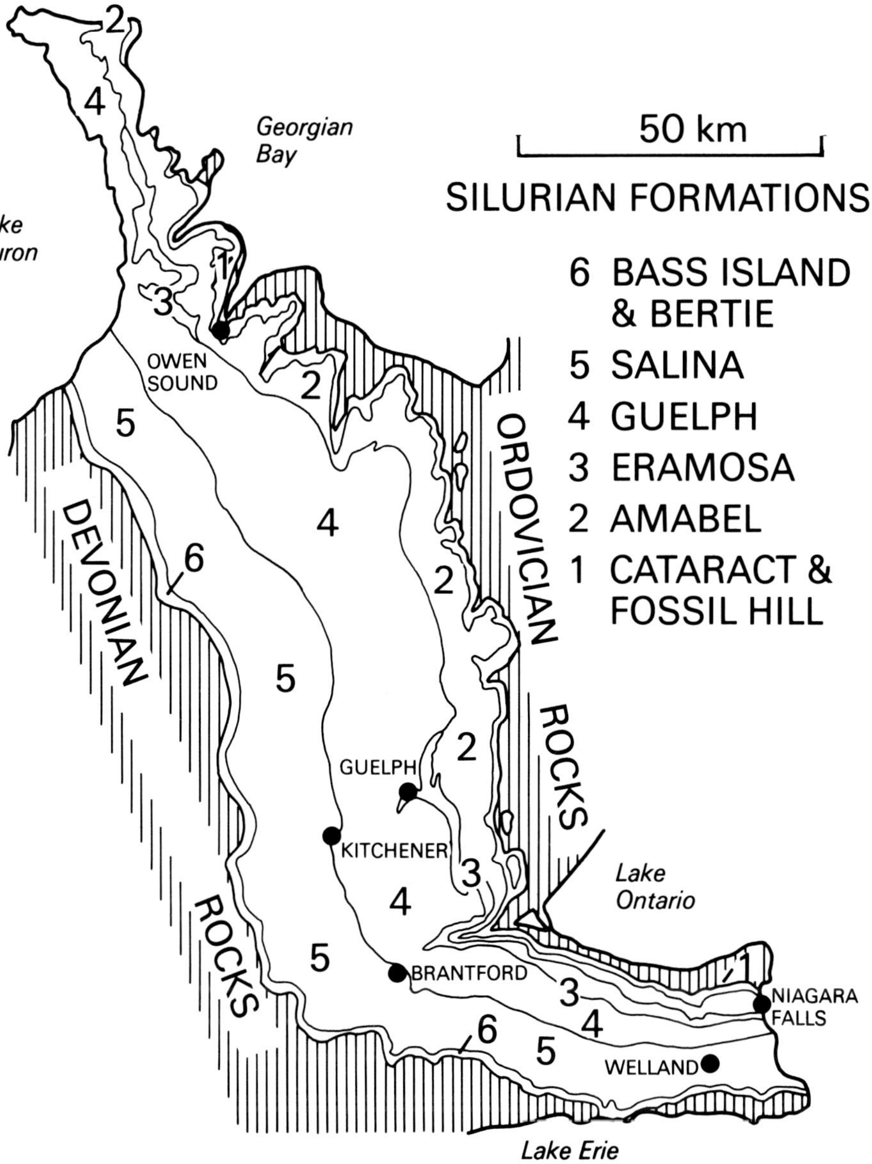

Fig. 4 Silurian geology of southern Ontario. Adapted from Winder and Sanford (1972).

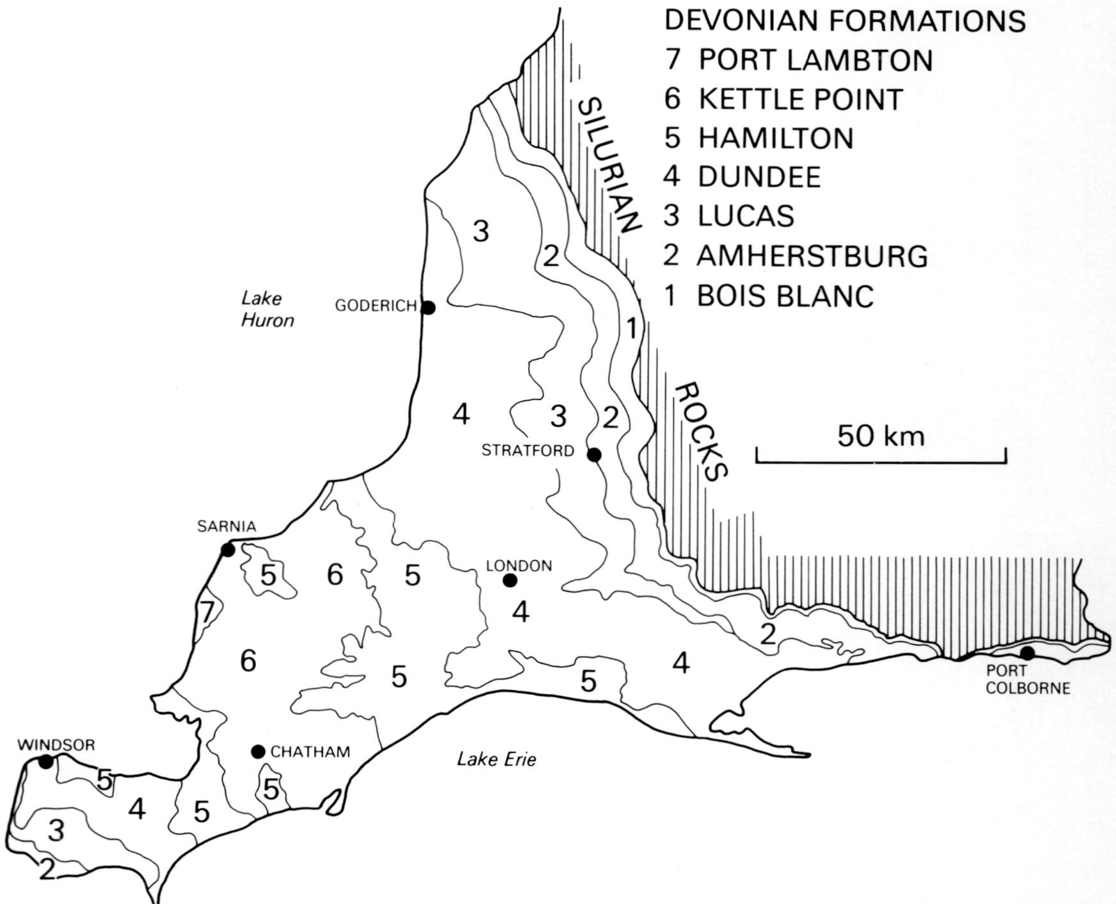

Fig. 5 Devonian geology of southern Ontario. Adapted from Winder and Sanford (1972).

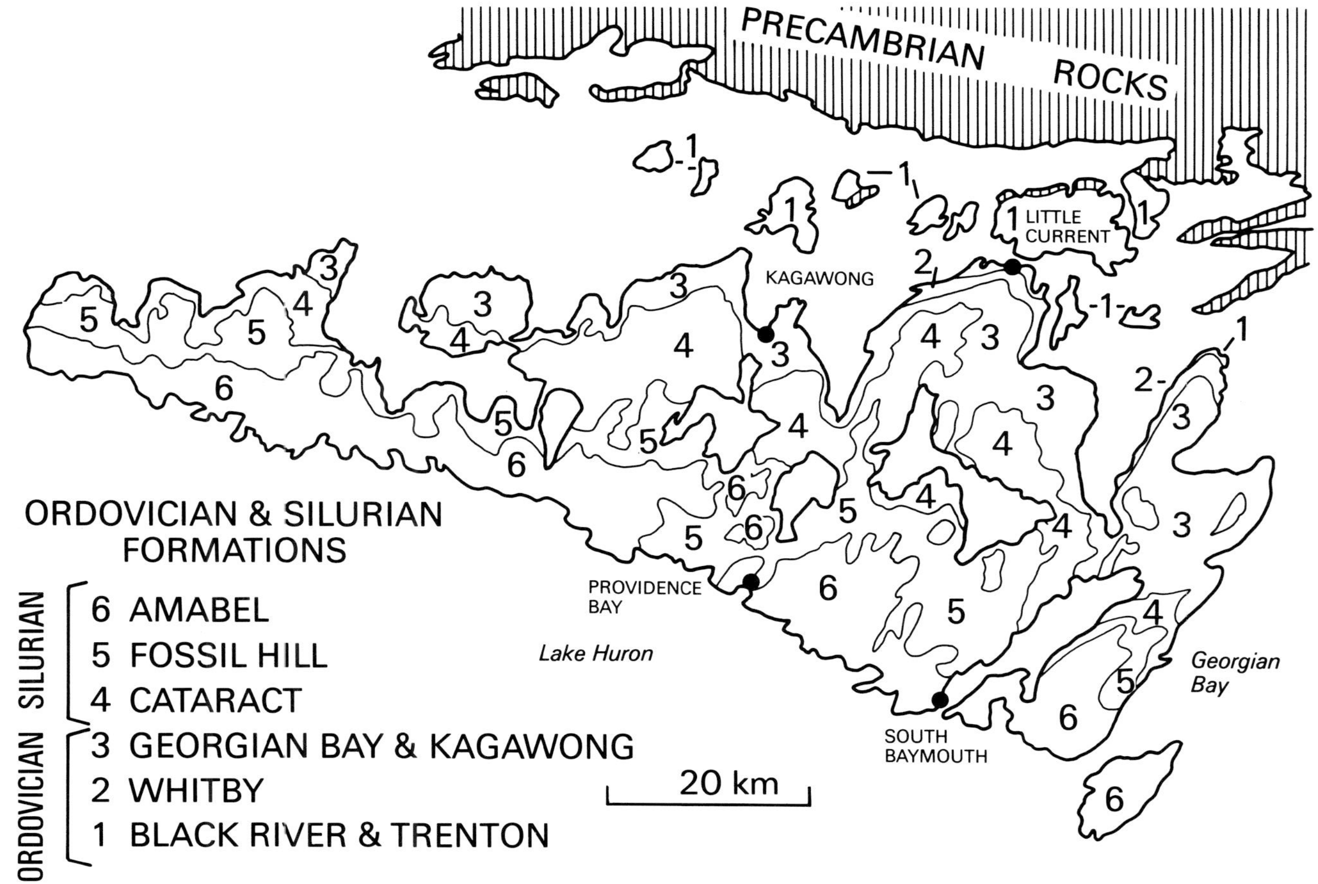

Fig. 6 Ordovician and Silurian geology of Manitoulin Island. Adapted from Liberty (1957).

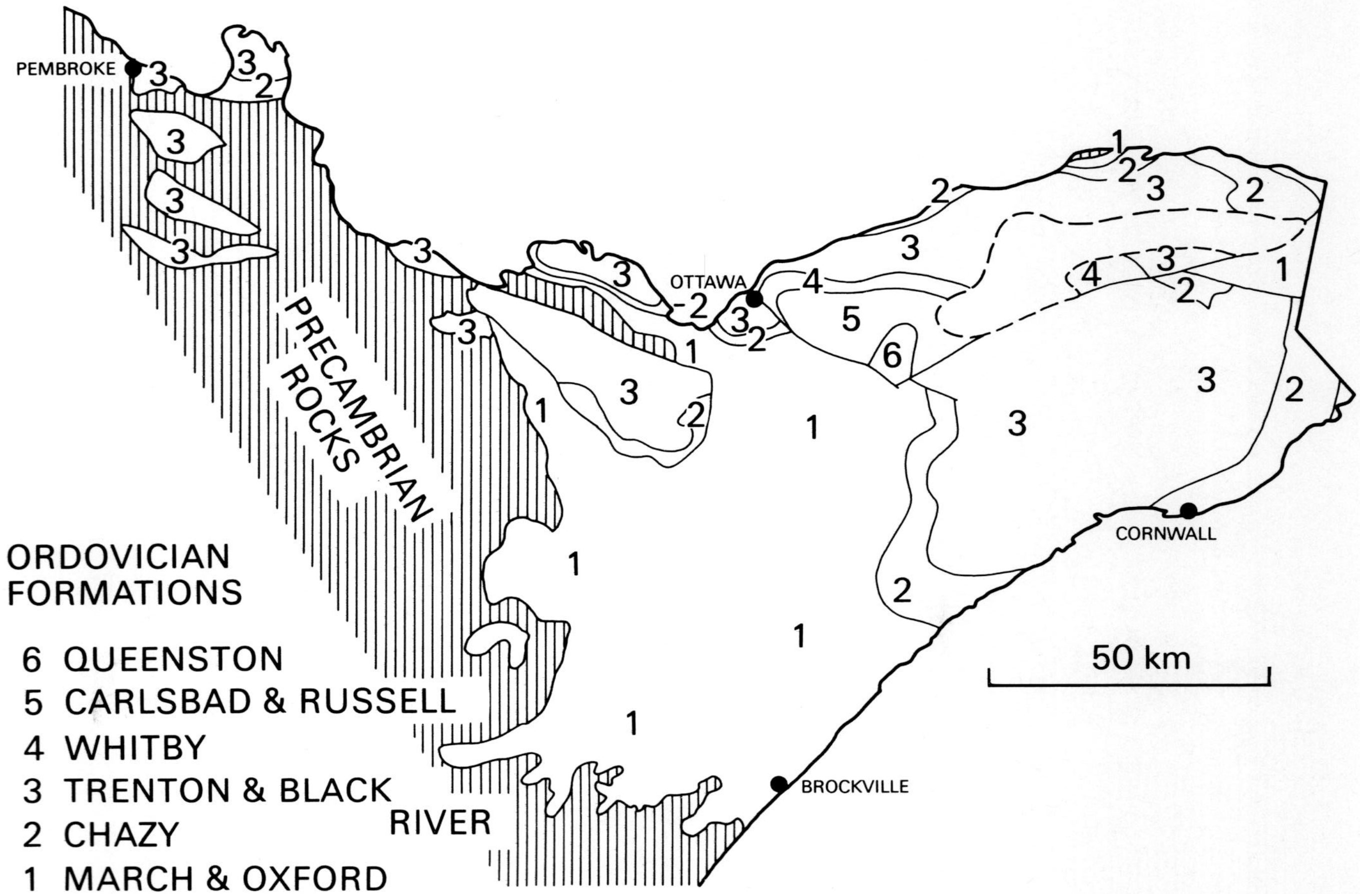

Fig. 7 Ordovician geology of the Ottawa area. Adapted from Hewitt (1972).

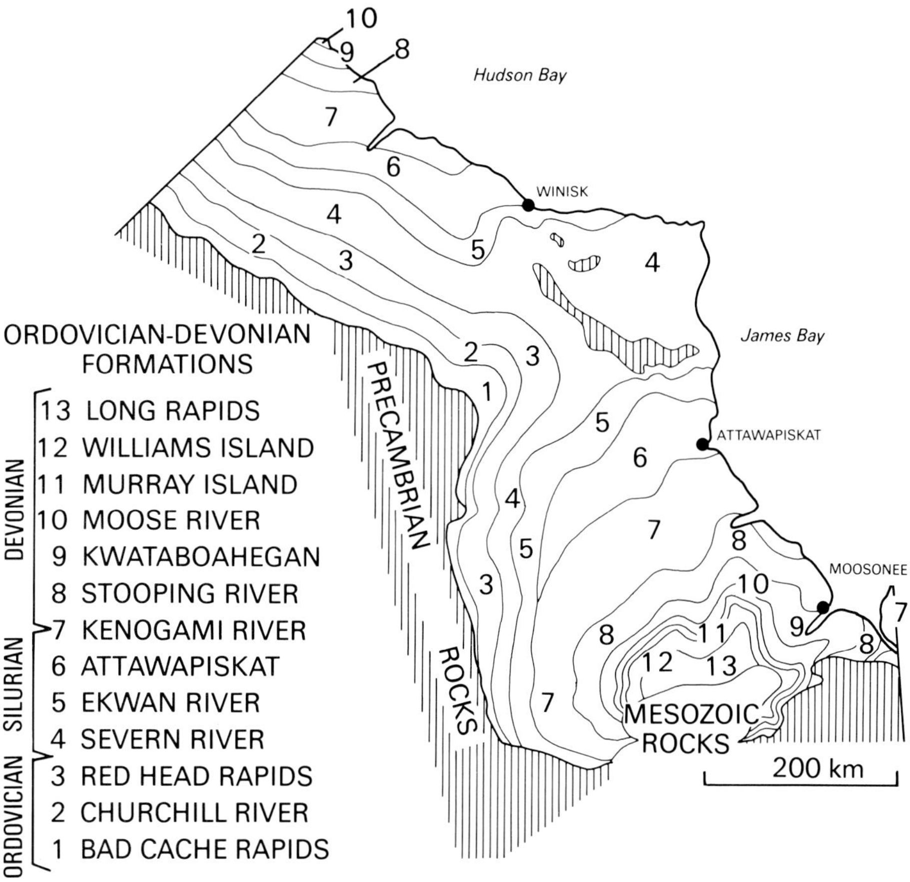

Fig. 8 Ordovician, Silurian, and Devonian geology of the Hudson Bay area. Adapted from Sanford and Norris (1973).

Palaeozoic Geology of Ontario

The sequence of Ordovician, Silurian, and Devonian formations in southern and northern Ontario is shown in Figure 2 and their outcrop patterns summarized in Figure 1 and Figures 3 to 8. The southern and northern areas of Palaeozoic outcrop are separated by a broad expanse of Precambrian igneous and metamorphic crystalline rocks on which occur scattered Palaeozoic outliers; the largest of these outliers is in the Lake Timiskaming area.

Because even a brief discussion of the Palaeozoic stratigraphy of Ontario is not possible here, references are given below.

GENERAL REFERENCES

Palaeozoic stratigraphy	
Ontario	Douglas (1970)
Southern Ontario	Winder and Sanford (1972)
	Winder *et al.* (1975)
	Hewitt (1972)
Hudson Bay area	Sanford and Norris (1973)
Ordovician stratigraphy	
Hudson Bay area	Cumming (1975)
Ottawa area	Wilson (1946)
Southern Ontario	Wilson *et al.* (1936)
	Liberty (1964)
Silurian stratigraphy	
Southern Ontario	Bolton (1957)
Hamilton area	Segall (1972)
Devonian stratigraphy	
Southern Ontario	Sanford (1968)
Hudson Bay area	Sanford and Norris (1975)

SPECIFIC REFERENCES

Toronto-Hamilton area	Caley (1940)
Brantford area	Caley (1941)
London area	Caley (1943)
Windsor-Sarnia area	Sanford and Brady (1955)
Lake Simcoe area	Liberty (1969)
Bruce Peninsula	Liberty and Bolton (1971)
Manitoulin Island	Liberty and Shelden (1968)
Lake Timiskaming	Bolton and Copeland (1972)

It is beyond the scope of this book to provide a detailed list of collecting localities in Ontario where specific trilobites may be obtained. Such a list would run into hundreds of entries. In any case, virtually all roadcuts, natural exposures, and quarries in the Palaeozoic rocks of Ontario are likely trilobite prospecting areas. Many guidebooks describe and locate collecting areas. The reader should consult Bolton (1957), Bolton and Copeland (1972), Liberty (1969), Liberty and Bolton (1971), Liberty and Shelden (1968), Segall (1972), Stauffer (1915), Stumm and Wright (1958), Wilson (1947), Winder *et al.* (1975), and Winder and Sanford (1972).

Introduction to Trilobites

Trilobites appeared abruptly at the beginning of the Cambrian and radiated rapidly to a diversity peak in the Late Cambrian. Their numbers declined markedly through the Ordovician, had a period of stability that lasted from the Early Silurian to the Middle Devonian, then another rapid decline to a period of slow attrition in numbers from the Late Devonian, through the Carboniferous, to their extinction in the Late Permian. This extinction involved only a handful of genera, and is a relatively insignificant event in light of the previous history of the group (Fig. 9). About 1,500 genera of trilobites existed during the one-third of a billion year history of the group.

The name trilobite refers to the longitudinal division of the body into an axial lobe and two pleural lobes. The body also shows a transverse division into a head region (cephalon), an articulating thorax, and a tail region (pygidium). The cephalon is generally crescentic in outline and comprises a central glabella outlined by axial and preglabellar furrows and flanked by a pair of cheeks. The glabella is usually inflated and may be divided into transverse lobes by an occipital furrow and a number of lateral glabellar furrows. In sighted trilobites, a compound eye is situated on each of the cheeks. The posterolateral portions of the cephalon may be blunt corners or may bear genal spines. Narrow lines of weakness where the cephalon splits during moulting are called facial sutures and these divide the cheek into an inner fixed cheek and an outer free cheek. The sutural pattern has been used as a classification criterion. These sutures may be opisthoparian (posterior part of suture cuts posterior margin of cephalon inside the genal angle; for example, *Pseudogygites*, Fig. 10), or proparian (posterior part of suture cuts lateral margin of cephalon in front of genal angle; for example, *Ceraurinella*, Fig. 11), or marginal (suture runs along margin of cephalon; for example, *Paraharpes*, Fig. 30A).

On the ventral side of the cephalon, the recurved portion of the dorsal exoskeleton or doublure, extends inwards for some distance. The doublure may be entire, or it may be traversed by a single median suture, or it may include a rostral plate outlined by sutures on all sides. Beneath the glabella occurs the hypostome; this plate is generally attached to the doublure with a sutural contact and appears to have covered the stomach and mouth regions of the trilobite.

The thorax consists of a number of bandlike segments between which articulation occurred. The distal portions of the thoracic segments may terminate in spines or may be faceted so that they can nest closely upon enrollment.

The pygidium consists of a single plate of fused segments that covers the pos-

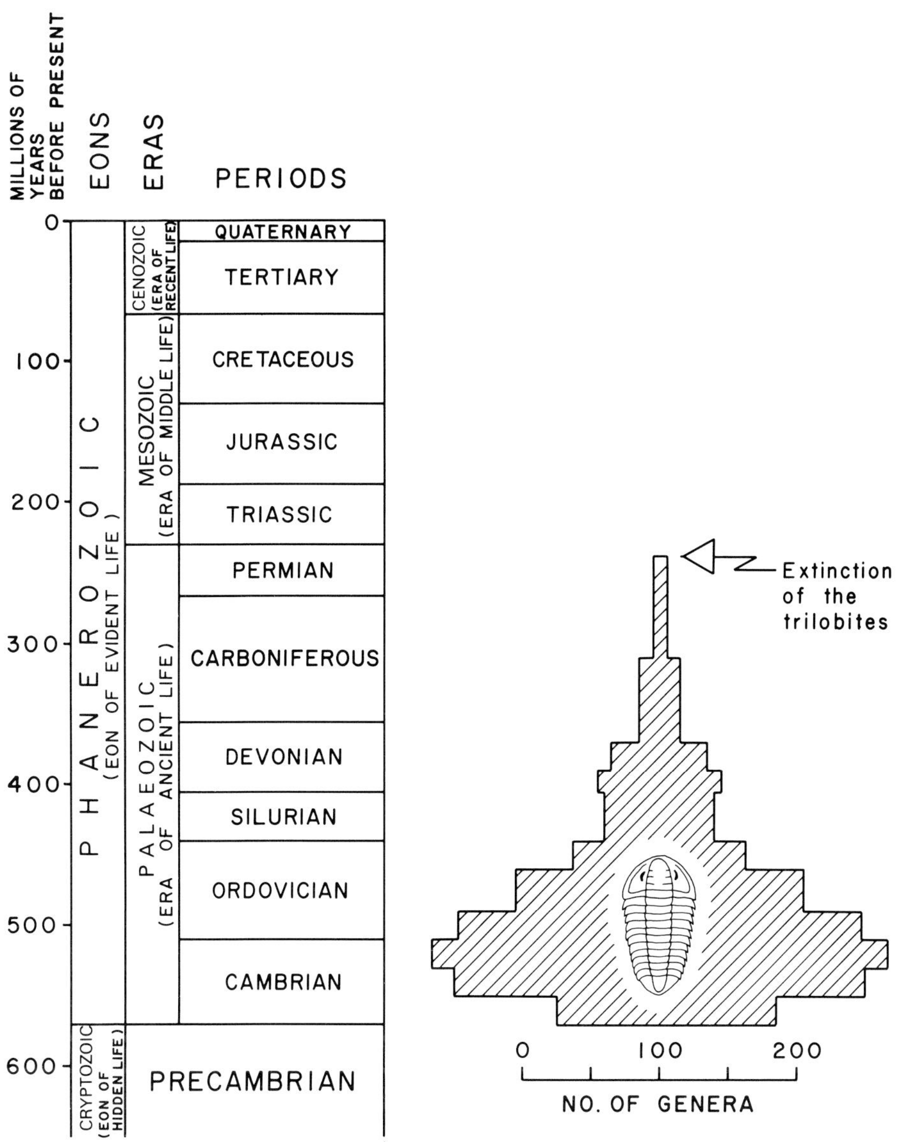

Fig. 9 Distribution of trilobites through time from their origin near the base of the Cambrian to their extinction in the late Permian. Data on trilobite generic diversity compiled from Moore (1959).

terior part of the trilobite. Its axis is composed of a number of axial rings; the axis is flanked by pleural portions traversed by pleural and interpleural furrows. Segmentation is not evident on all pygidia; some are perfectly smooth. The shape and size of pygidia vary widely and may bear a number of large or small marginal spines.

Trilobites that reveal morphological details other than those of the hard exoskeleton are rarely found. Information about the anatomy of trilobites is thus obtained only from study of exceptionally preserved specimens such as those il-

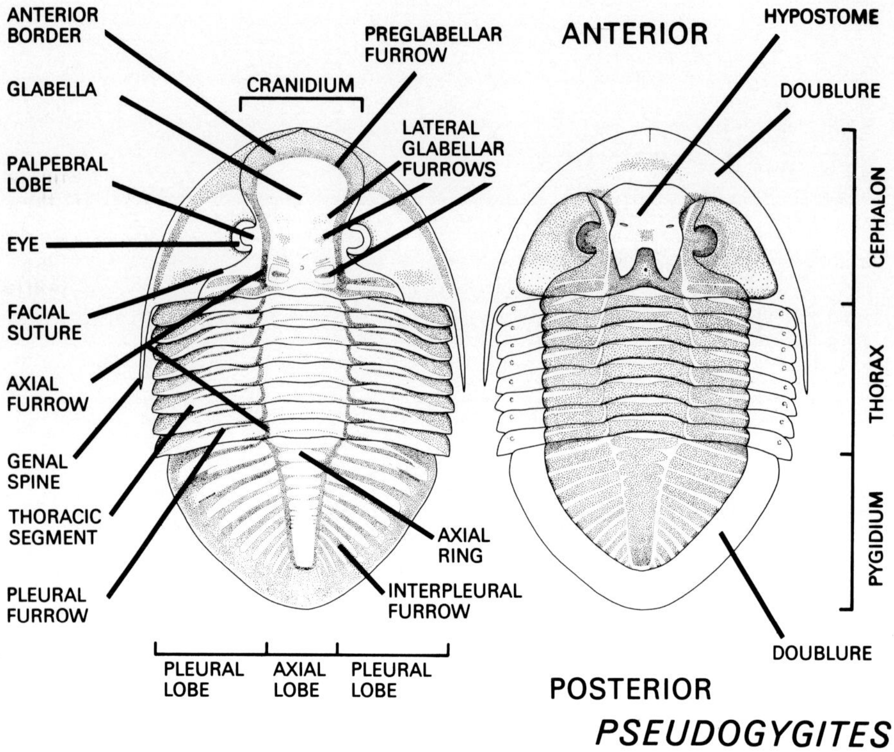

Fig. 10 Morphologic terms applied to an opisthoparian trilobite (*Pseudogygites*). Dorsal (left) and ventral (right) views.

lustrated in Figure 12. The exoskeleton, the appendages, and the internal features of these trilobites have been replaced by finely crystalline pyrite (FeS_2). The colour contrast of the pyrite against the matrix may be used for photography by immersing the trilobite in a liquid of high refractive index (Levi-Setti, 1975). The pyrite replacement also enables investigation by radiographic (X-ray) techniques (Cisne, 1975). These and other similar techniques show that the appendages of trilobites consisted of a sequence of paired sets of bladelike bars (coxa) located under the axis. Emerging in a lateral direction from each of these bars is a dorsal featherlike gill branch (exite) and a ventral jointed walking leg (telepod). A pair of antennae emerges from the inner surface of the hypostome. Fine details of the muscular, digestive, circulatory, visual, feeding, and locomotory systems of trilobites have been studied in similarly preserved fossils (Stürmer and Bergström, 1973; Cisne, 1975; Whittington, 1975; Campbell, 1975a).

Like other arthropods, trilobites moulted periodically during growth. Therefore, the growth history of an individual trilobite is not recorded as growth lines or growth rings (as is the case in organisms with accretionary skeletons such as trees, clams, and humans), but instead the life history (or ontogeny) of a trilobite is recorded in a series of moulted exoskeletons that show progressive size increase and shape change. The earliest exoskeleton is a minute undivided oval

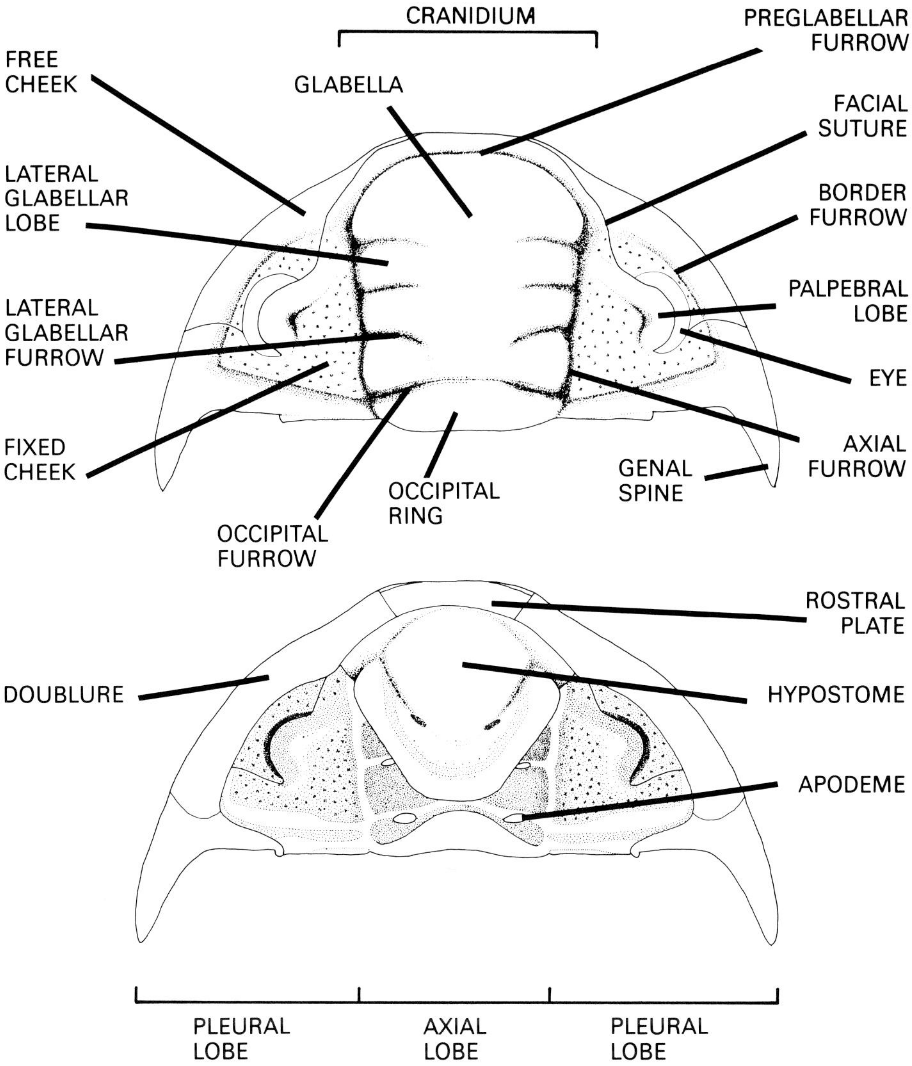

Fig. 11 Morphologic terms applied to the cephalon of a proparian trilobite (*Ceraurinella*). Dorsal (upper) and ventral (lower) views.

shield assigned to the protaspid stage (Fig. 13A). The meraspid stage (Fig. 13B) is initiated by the appearance of a joint between the cephalon and the pygidial portion. Successive meraspid exoskeletons contain increased numbers of thoracic segments (Fig. 13C–G) until the full complement of thoracic segments for that species is attained; this moult delineates the holaspid stage (Fig. 13H). The trilobite continues to moult and grow within the holaspid stage, but no additional segments are added to the thorax.

Many trilobite specimens do not record the death of an individual, but

Fig. 12 Dorsal (left) and ventral (right) views of two pyritized specimens of *Triarthrus eatoni* (Hall) showing the hard part morphology and the appendages. Approximately × 2.5. The specimens are from the Utica Shale (Upper Ordovician), near Rome, New York State and were collected and prepared by C.E. Beecher during the 1890s. The photographs are by Prof. Levi-Setti of the University of Chicago who captured the fine detail and contrast by immersing the specimens in xylene. Reprinted from *Trilobites: a photographic atlas* by R. Levi-Setti (1975, pls. 7 and 9) with permission of the University of Chicago Press.

merely represent moults of a living trilobite. Therefore, a single trilobite may leave as many complete fossils as it had moult stages. The moult remains were generally scattered across the sea floor by currents. Occasionally, moult remains occur in the same configuration as the trilobite left them: a good example of this from the shales of the lower Whitby Formation at Craigleith is shown in Figure 14. If we compare the disposition of the moulted elements of *Pseudogygites latimarginatus* (Fig. 14B) with a dead individual of the same species (Fig. 14A), we can reconstruct the behaviour of the moulting trilobite in a series of diagrams (Fig. 14C).

One of the fascinating aspects of trilobites is the presence in many species of well-developed paired eyes; and because this visual system is the earliest about which information is available, it attracts considerable attention. The optical medium was precisely oriented crystals of calcium carbonate and, as such, the trilobite eyes are unique in the animal kingdom. Excellent accounts of trilobite eyes and vision are given by Towe (1973), Clarkson (1975), Clarkson and Levi-

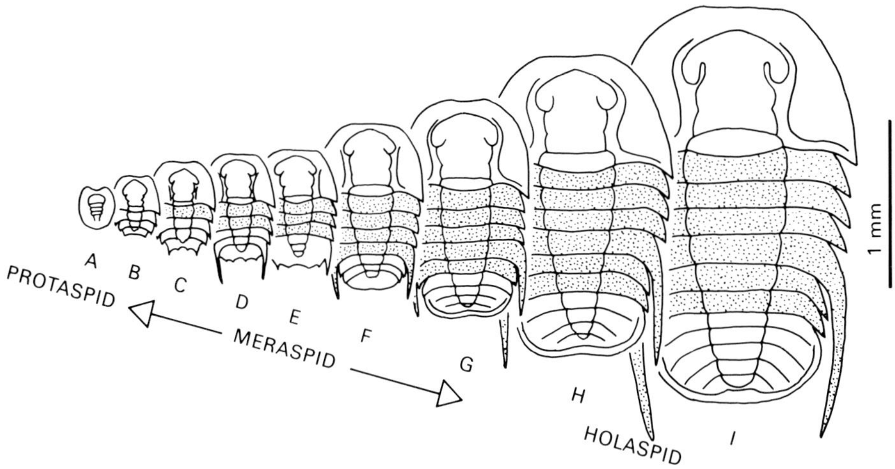

Fig. 13 Life history of a trilobite, *Shumardia pusilla* (Sars), showing the division into the larval stage (protaspid, A), the juvenile stage (meraspid, B–G), and the adult stage (holaspid, H and I). Note the increase in the number of thoracic segments through the meraspid stage from 0 to 5. The holaspid stage is defined as when the full number of thoracic segments is reached—in this case, 6. Redrawn from Stubblefield (1926).

Setti (1975), and Levi-Setti (1976). Two types of trilobite eye have been defined: the holochroal eye (Figs. 21D, 24E) consists of a visual surface of numerous, small, densely packed lenses covered by a single continuous cornea. The schizochroal eye (Figs. 24B, 56, 57B) consists of a visual surface of large separated lenses, each of which is covered by a single cornea. The holochroal eye is, by far, the most common trilobite eye and occurs in all groups except the phacopids and dalmanitids which have schizochroal eyes.

Most sighted trilobites appear to have been benthic (bottom dwelling) animals that lived in generally shallow and well-lit marine waters. Many, particularly those with large eyes and ornamented exoskeletons, were probably epifaunal (lived on the sediment or furrowed through the top few millimetres of sediment). In such settings, trilobites could have been filter feeders, detritus feeders, scavengers, or even predators. Others, particularly those large smooth forms with broad axes such as *Bumastoides, Bumastus, Anataphrus*, and *Dipleura*, were probably semi-infaunal (lived partially submerged in the sediment) and were either detritus or filter feeders. Trilobites with reduced pleural portions and downwardly directed eyes may have been nektic (swimmers) or planktic (floaters). Blind trilobites could have been deep burrowers or could have lived in habitats below the photic zone.

Undoubtedly the best and most authoritative source book on trilobites is the *Treatise* (Moore, 1959). This volume summarizes available knowledge on trilobites and includes diagnoses and illustrations of most genera described up to 20 years ago. Information about trilobites and other fossils may be found in introductory palaeontology textbooks such as Moore *et al.* (1952), and Black (1970) and in nontechnical books such as MacFall and Wollin (1972). Finally, *Trilobites: a photographic atlas* by R. Levi-Setti (1975) is highly recommended as a superbly illustrated and well-written passport into the world of trilobites.

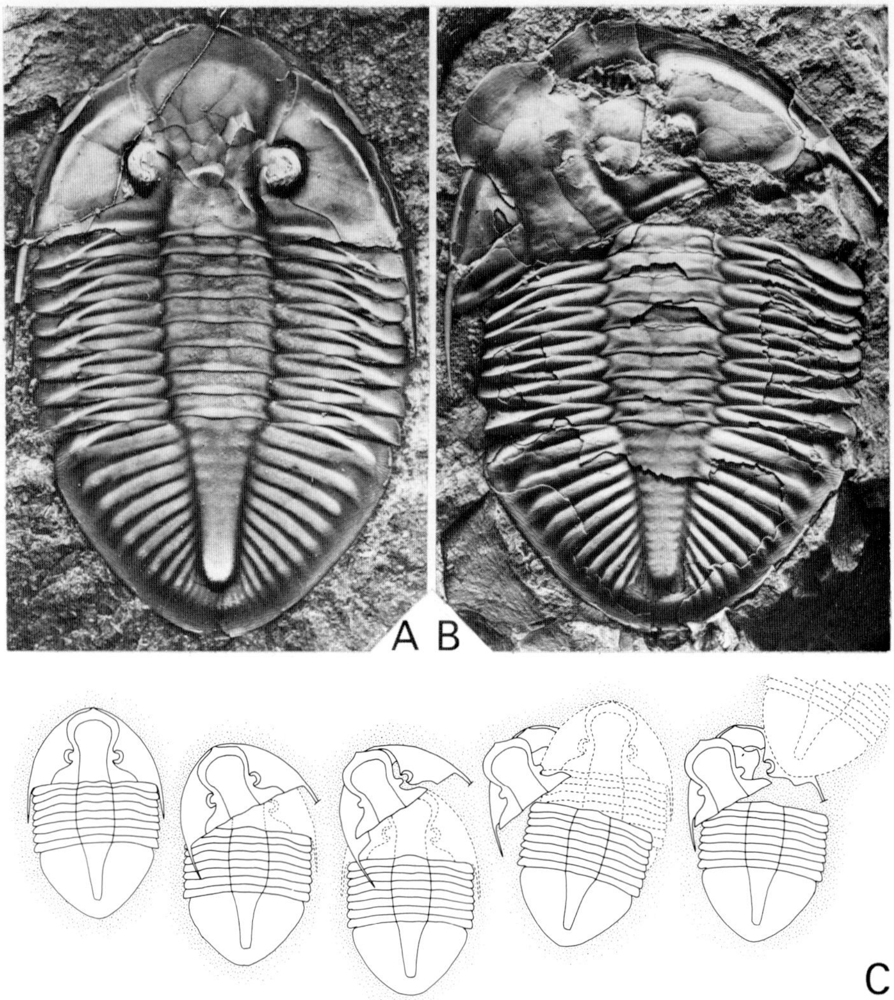

Fig. 14 How the trilobite *Pseudogygites* moulted.

A A dead individual (sutures essentially closed, hypostome in place), ROM 35029, × 2.5.

B The moult remains left by the trilobite, ROM 30015, × 1.8.

C A moult scenario.

The trilobite severed the connection between the thorax and cephalon by anchoring one genal spine in the mud and backing up. This caused the cephalon to rotate back over the front part of the thorax. With its head, the trilobite then nudged the cranidium free from the remaining cephalon and crawled obliquely forward through the gap left by the split facial suture. As the trilobite moved forward it pushed the cranidium while the conjoined thorax and pygidium became caught against the cephalon and slid off the back of the trilobite. Finally, when it was almost free of its moult remains, the trilobite dislodged and overturned the hypostome with its thrashing legs.

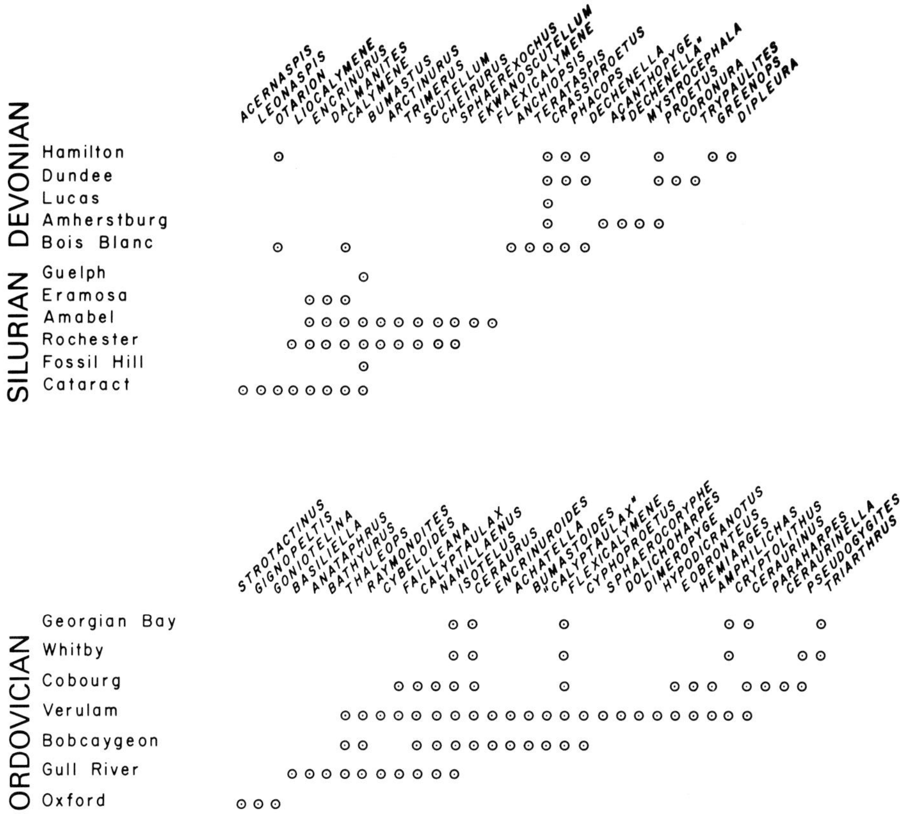

Fig. 15 Stratigraphic occurrences of Ordovician, Silurian, and Devonian trilobite genera in southern Ontario.

History of Trilobite Studies in Ontario

In 1820, Dr. John J. Bigsby (Fig. 16A), at that time the British representative to the Boundary Commission between British North America and the United States, collected fossils from limestones exposed on St. Joseph Island in Lake Huron. Among the fossils were several trilobite specimens and, on his return to England, Bigsby gave them to Charles Stokes. In an appendix to Bigsby's paper on the geography and geology of Lake Huron published in 1824, Stokes illustrated and briefly commented on this trilobite and gave it a name—*Asaphus platycephalus*. This was the first trilobite from the area now included in Ontario to receive a name, and was also the first trilobite from North America to be described. In addition, Stokes' note was the first published identification of the plate that covers the mouth region in trilobites, the hypostome. The original illustrations of this trilobite, now identified as *Isotelus platycephalus* (Stokes), are shown in Figure 17 along with photographs of the specimen in its present state.

Little further work was done on Ontario trilobites until 1856 when Sir William Logan appointed Elkanah Billings (Fig. 16B) palaeontologist to the Geological Survey of Canada. Billings developed a deep interest in natural history and fossils while pursuing twin careers in law and as the editor of the *Bytown Citizen*. He also founded and served as the editor and chief contributor to a scientific journal called *The Canadian Naturalist and Geologist*. After 1856 Billings started the massive task of arranging, identifying, and classifying the fossils collected by members of the Survey from all parts of Upper and Lower Canada. These fossils, including trilobites, were described in a long series of papers published during the next 20 years. Billings' authoritative initiation of Canadian Palaeozoic palaeontology is perhaps best displayed in his contribution to Logan's monumental *Geology of Canada* published in 1863 and in his *Palaeozoic Fossils* published in 1865. Elkanah Billings died in 1876 and is buried in Toronto.

Contemporaries of Billings also involved in the study of Ontario trilobites include James Hall, State Palaeontologist of New York, whose work on Silurian and Devonian trilobites extended into Ontario, and E.J. Chapman, Professor of Mineralogy and Geology at the University of Toronto. Chapman published an obsolete classification of trilobites in 1890.

The next major personage in Ontario trilobite studies was Percy E. Raymond (Fig. 16C) who was Chief Palaeontologist to the Geological Survey of Canada from 1910 to 1912. Raymond's particular forte was Ordovician trilobites and he produced a large number of papers on trilobites of the Trenton and Black River Groups in southern Ontario.

The only comprehensive work on trilobites of a specific region in Ontario was that of Alice E. Wilson (Fig. 16D) in 1947. This book was on Middle Ordovician trilobites of the Ottawa area. It formed part of an extensive series of monographs on the Ordovician invertebrate fossils of the region published between 1945 and 1961.

More recently other investigators have contributed towards an understanding of Ontario trilobites: G.W. Sinclair of the Geological Survey of Canada on Ordovician trilobites, E.C. Stumm of the University of Michigan on Devonian trilobites, and H.B. Whittington of Harvard University on Ordovician, Silurian, and Devonian trilobites. Their contributions are in the reference list.

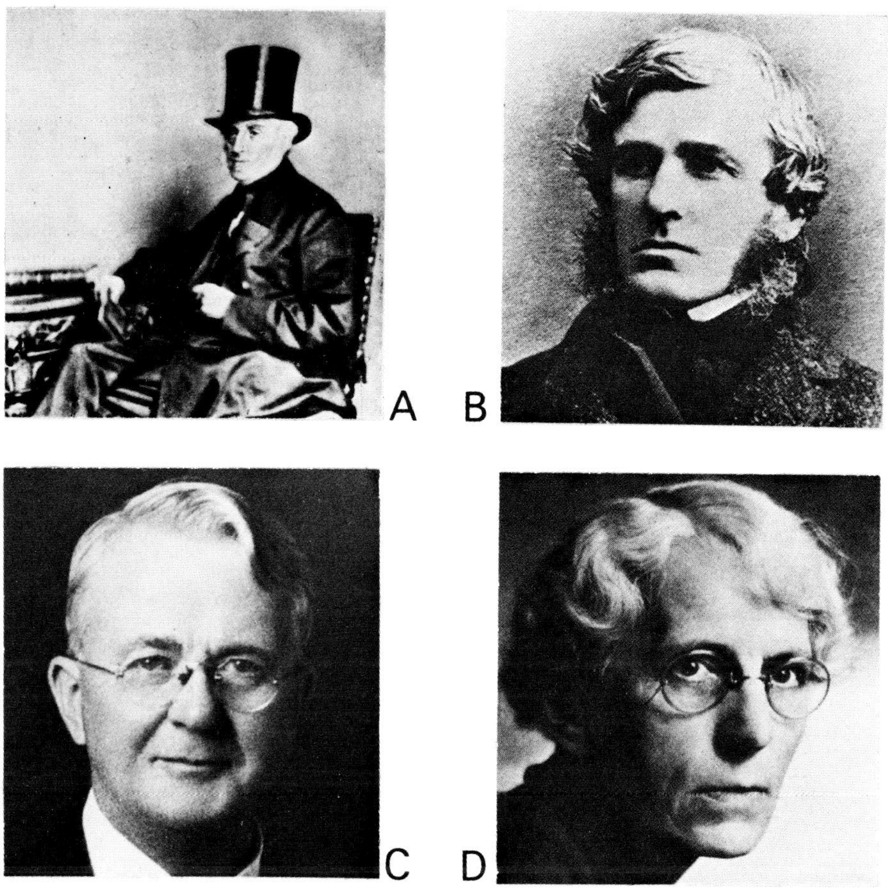

Fig. 16 Students of Ontario trilobites.
- A John J. Bigsby, 1792–1881
- B Elkanah Billings, 1820–1876
- C Percy E. Raymond, 1879–1952
- D Alice E. Wilson, 1881–1964

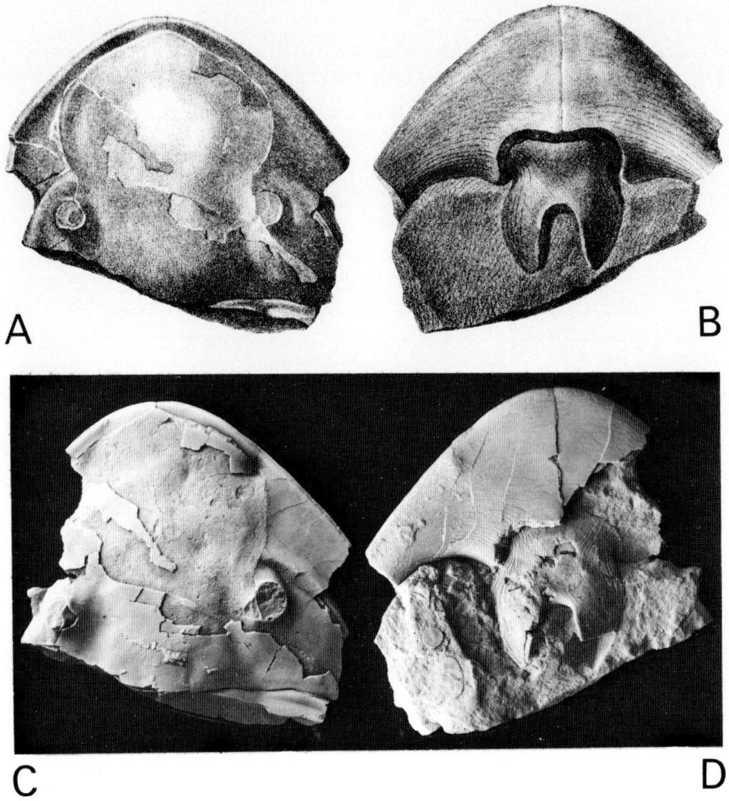

Fig. 17 *Isotelus platycephalus* (Stokes)—the first Ontario trilobite to be described. The specimen was collected in 1820 by Dr. J.J. Bigsby on St. Joseph Island.

A,B The illustration of *Asaphus platycephala* published by Charles Stokes (in Bigsby, 1824). Dorsal and ventral views of cephalon, × .65.

C,D The specimen as it appears now, BM I82, × .65.

Types of Trilobites Represented in Ontario

I will not provide a formal classification of trilobites; the reader is referred to the *Treatise* (Moore, 1959) and to Bergström (1973) for two differing interpretations of trilobite classification. I will merely present outline drawings of a typical member of each of 21 informal groupings of genera that occur in Ontario (Fig. 18), indicate (in brackets) the corresponding classification according to the *Treatise*, and list the Ontario genera that belong in each.

Olenid trilobites (Family Olenidae)
Triarthrus
Asaphid trilobites (Family Asaphidae)
Isotelus, Basiliella, Anataphrus, Pseudogygites
Remopleuridid trilobites (Family Remopleurididae)
Hypodicranotus
Illaenid trilobites (Family Illaenidae)
Failleana, Nanillaenus, Thaleops, Bumastoides, Bumastus
Scutelluid trilobites (Family Thysanopeltidae)
Eobronteus, Scutellum, Ekwanoscutellum
Bathyurid trilobites (Superfamily Bathyuracea)
Bathyurus, Raymondites, Goniotelina, Gignopeltis
Dimeropygid trilobites (Family Dimeropygidae)
Dimeropyge
Proetid trilobites (Family Proetidae)
Crassiproetus, Proetus, Cyphoproetus, Dechenella, "Dechenella"
Otarionid trilobites (Family Otarionidae)
Otarion
Brachymetopid trilobites (Family Brachymetopidae)
Mystrocephala
Harpid trilobites (Family Harpidae)
Paraharpes, Dolichoharpes
Trinucleid trilobites (Family Trinucleidae)
Cryptolithus
Cheirurid trilobites (Family Cheiruridae)
Ceraurus, Ceraurinus, Ceraurinella, Sphaerocoryphe, Cheirurus, Sphaerexochus
Pliomerid trilobites (Family Pliomeridae)
Strotactinus

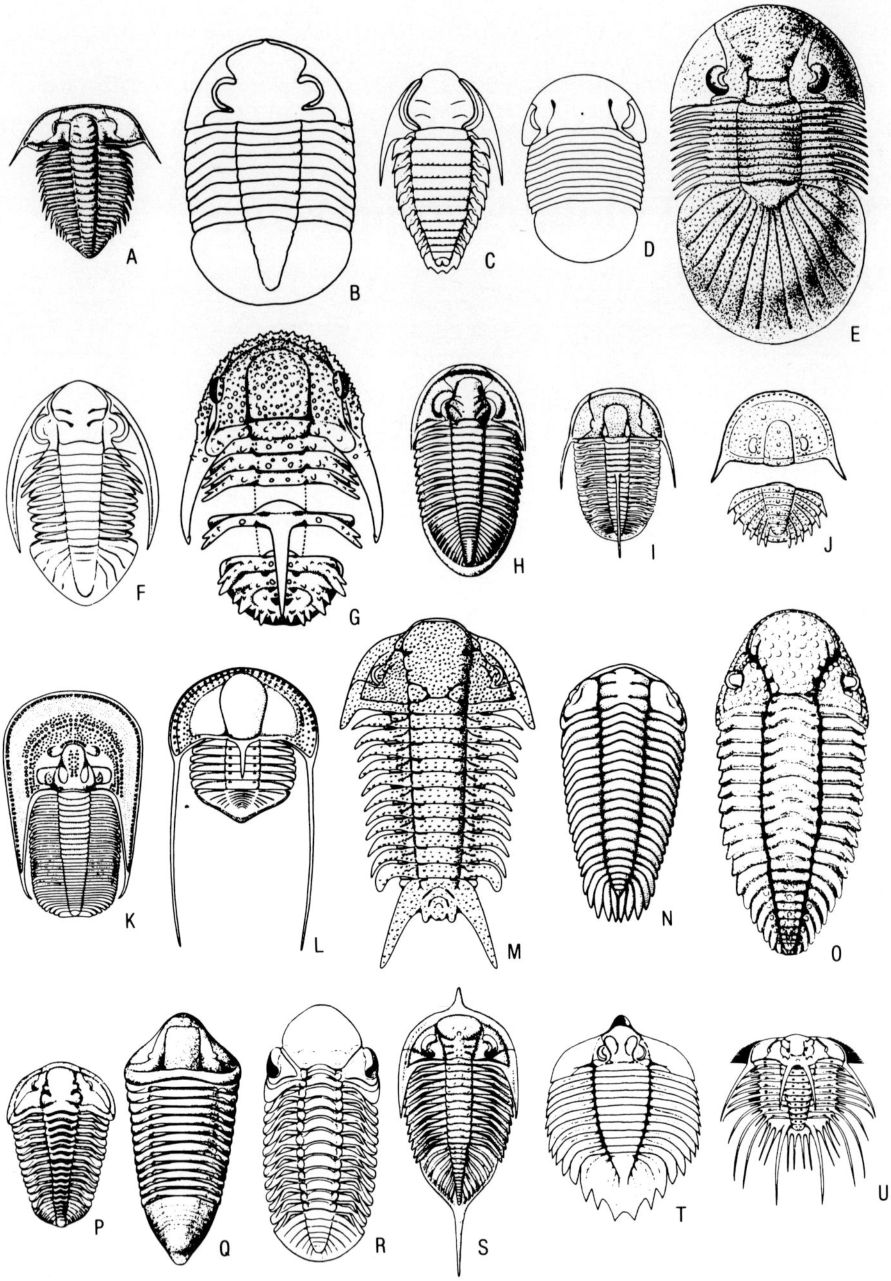

Fig. 18 Types of trilobites represented in Ontario. Mainly culled from Moore (1959). Not to scale.

A Olenid (*Olenus* illustrated)
B Asaphid (*Plectasaphus*)
C Remopleuridid (*Remopleurides*)
D Illaenid (*Bumastus*)
E Scutelluid (*Scutellum*)
F Bathyurid (*Bathyurus*)
G Dimeropygid (*Dimeropyge*)
H Proetid (*Dechenella*)
I Otarionid (*Otarion*)
J Brachymetopid (*Brachymetopus*)
K Harpid (*Harpes*)
L Trinucleid (*Cryptolithus*)
M Cheirurid (*Ceraurinella*)
N Pliomerid (*Pliomerops*)
O Encrinurid (*Encrinurus*)
P Calymenid (*Calymene*)
Q Homalonotid (*Dipleura*)
R Phacopid (*Reedops*)
S Dalmanitid (*Dalmanites*)
T Lichid (*Arctinurus*)
U Odontopleurid (*Odontopleura*)

Encrinurid trilobites (Family Encrinuridae)
Encrinurus, Encrinuroides, Cybeloides
Calymenid trilobites (Subfamily Calymeninae)
Flexicalymene, Liocalymene, Calymene
Homalonotid trilobites (Subfamily Homalonotinae)
Trimerus, Dipleura
Phacopid trilobites (Superfamily Phacopacea)
Acernaspis, Phacops
Dalmanitid trilobites (Superfamily Dalmanitacea)
Calyptaulax, Eomonorachus, "Calyptaulax", Achatella, Dalmanites, Anchiopsis, Greenops, Trypaulites, Coronura
Lichid trilobites (Order Lichida)
Hemiarges, Amphilichas, Dicranopeltis, Terataspis, Arctinurus, Acanthopyge
Odontopleurid trilobites (Order Odontopleurida)
Leonaspis

Ordovician Trilobites

Figs. 19 to 39

At least 34 genera of trilobites are found in Ordovician rocks of southern Ontario (Fig. 15). Trilobites are rare in the Lower Ordovician dolostones of the Ottawa area where only a few specimens are known. They are more diverse and abundant in the Black River Group, but are often difficult to collect because of the preponderance of tough, fine-grained limestones. Fossiliferous limestones of the Trenton Group have yielded abundant and well-preserved specimens and trilobites reach a diversity peak with 23 genera in the limestones and shales of the Verulam Formation. Even here, trilobites are not among the more conspicuous faunal elements. Of the 330 species of fossils recorded from the Verulam Formation (Liberty, 1969) in the Lake Simcoe area only 31 are trilobites. The only stratigraphic interval in Ontario in which trilobites occur on virtually every piece of rock and consistently outnumber all other fossils is the lower part of the Whitby Formation. Here, dark brown bituminous shales contain massed specimens of *Pseudogygites* and *Triarthrus*. Only a few poorly preserved trilobites occur in the younger Ordovician formations in southern Ontario. The very large specimens of *Isotelus* in the Georgian Bay Formation are rare exceptions.

Trilobites have not been described from the Ordovician limestones of the Hudson Bay area and only a few genera are recorded, including *Ceraurinus* and *Isotelus* in the Bad Cache Rapids Group and the same genera plus *Nanillaenus, Encrinuroides, Calyptaulax*, and *Bumastoides* in the overlying Churchill River Group.

***Strotactinus* Bradley Fig. 19A, B**
Strotactinus is one of the few trilobites that occurs in the Lower Ordovician dolostones of the Ottawa area. It has a convex cranidium and a rectangular glabella with three pairs of straight lateral furrows; the specimen illustrated in Figure 19A is an internal mould and only two pairs of lateral furrows are evident. The anterior border of the cranidium is high and roll-like and the eyes are located far out on the cheeks opposite the second glabellar lobe. The pygidium consists of a broad segmented axis flanked by five pairs of downwardly curving tubular ribs that terminate as short pointed spines.
SPECIES *Strotactinus salteri* (Billings)—Oxford Formation, Ottawa area.
REFERENCES Billings (1865), Bradley (1925)

***Goniotelina* Whittington and Ross Fig. 19C,D**

This early bathyurid trilobite has an inflated and parallel-sided glabella lacking furrows and large palpebral lobes with distinct curving palpebral furrows. The triangular pygidium has a well-defined axis bearing four furrows, concave borders, and a terminal spine. The surface of the cranidium and pygidium is sparsely granulate.

SPECIES *Goniotelina subrectus* (Bradley)—Oxford Formation, Ottawa area.

REFERENCE Whittington (1953)

***Gignopeltis* Raymond Fig. 19E**

Only a few pygidia of this genus are known and it is difficult to decide how it should be classified. The pygidium of *Gignopeltis* is sub-semicircular in outline, evenly inflated in cross-section, and lacks a border. The axis is outlined by shal-

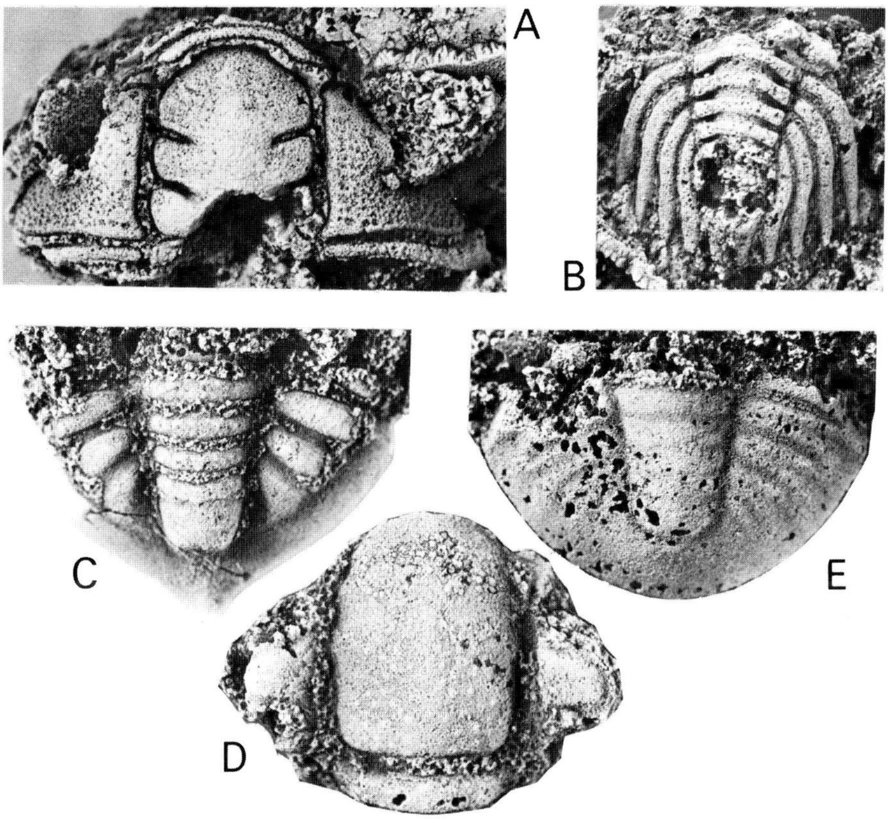

Fig. 19 ***Strotactinus Goniotelina Gignopeltis***
Strotactinus salteri (Billings)
A Cranidium, × 4.6, GSC 7834, Oxford Fm., Oxford Township.
B Pygidium, × 3.5, GSC 515, Oxford Fm., Oxford Township.
Goniotelina subrectus (Bradley)
C Pygidium, × 3.2, ROM 19a CAL, Oxford Fm., Marlborough Township.
D Cranidium, × 2.6, ROM 19b CAL, Oxford Fm., Marlborough Township.
Gignopeltis convexus (Billings)
E Pygidium, × 3.2, ROM 18 CAL, Oxford Fm., Marlborough Township.

low furrows and the pleural lobe is crossed by four faint furrows. Such a pygidium is not unlike those of some bathyurid trilobites.

SPECIES *Gignopeltis rarus* (Billings) and *Gignopeltis convexus* (Billings)—Oxford Formation, Ottawa area.

REFERENCES Billings (1865), Raymond (1924)

***Basiliella* Kobayashi Fig. 20A,B**

Basiliella is an asaphid trilobite characterized by the markedly divergent course of the facial suture in front of the large eyes, the very long and broad genal spines, and the strongly ribbed pygidium. Like many other asaphid genera, *Basiliella* has a forked hypostome.

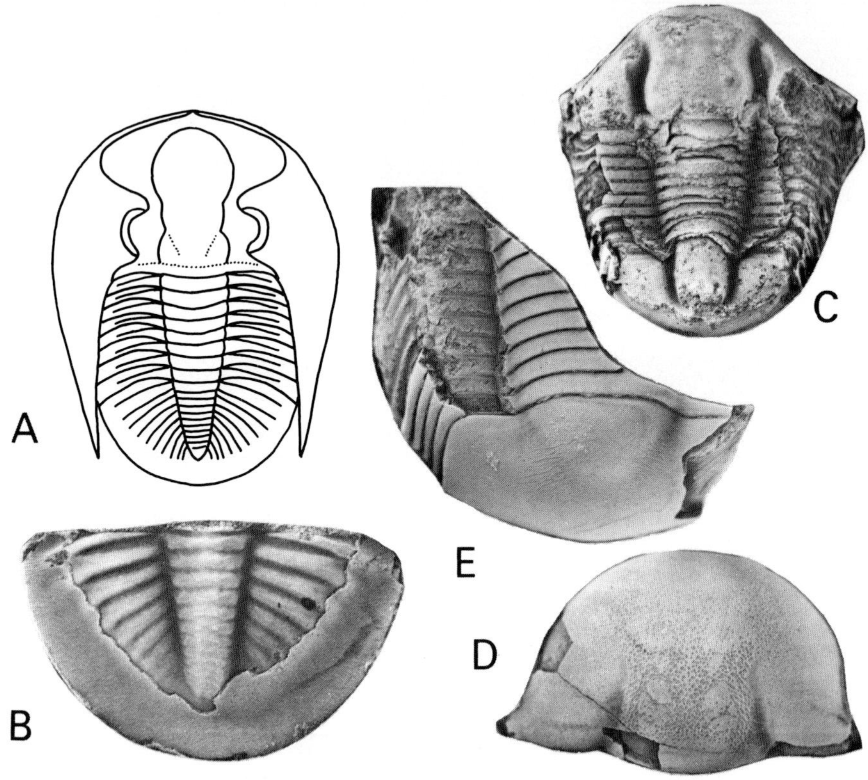

Fig. 20 ***Basiliella Thaleops Nanillaenus***
Basiliella barrandei (Hall)
A Reconstruction, × .9.
B Pygidium, × 2.1, ROM 18695, Black River Gp., Ottawa.
Thaleops ovata Conrad
C Exoskeleton, × 1.6, GSC 1321, Black River Gp., Ottawa.
Nanillaenus latiaxiatus (Raymond and Narraway)
D Cranidium, × 1.7, ROM 18754a, Black River Gp., Ottawa.
E Thorax and pygidium, × 1.7, ROM 18754b, Black River Gp., Ottawa.

SPECIES *Basiliella barrandei* (Hall)—Black River Group, southern Ontario and Ottawa area.
REFERENCE Raymond (1916)

Thaleops Conrad Fig. 20C
This illaenid trilobite has an axis defined by deep furrows. These axial furrows are nearly obsolescent on the posterior part of the pygidium and the anterior part of the cephalon. The eyes of *Thaleops* are elevated and the genal spines are fairly long and hooked inwardly.
SPECIES *Thaleops ovata* Conrad—upper Black River and Trenton Groups, southern Ontario and Ottawa area.
REFERENCES Wilson (1947), Jaanusson (1954)

Nanillaenus Jaanusson Fig. 20D,E
Nanillaenus is rather similar to *Thaleops*; the principal difference is the presence of shallower axial furrows, less elevated eyes, and shorter genal spines.
SPECIES *Nanillaenus conradi* (Billings), *Nanillaenus latiaxiatus* (Raymond and Narraway), *Nanillaenus angusticollis* (Billings), *Nanillaenus americanus* (Billings) —upper Black River and Trenton Groups, southern Ontario and Ottawa area.
REFERENCES Billings (1865), Sinclair (1947), Wilson (1947), Jaanusson (1954), Chatterton and Ludvigsen (1976)

Bathyurus Billings Fig. 21
Bathyurus has a large inflated glabella with two pairs of faint lateral glabellar furrows. The palpebral lobes are large and semicircular; the genal spines are moderately long and tapering. The pygidium is triangular and has four interpleural furrows. The square hypostome is diagnostic, as is the pecular rostral plate. This plate has not yet been recognized on specimens from Ontario, but silicified material from the Northwest Territories shows it to perfection (Fig. 21E). *Bathyurus* is moderately common in the fine-grained limestones of the Black River Group in southern Ontario. It probably lived partially submerged in the carbonate mud with only the eyes protruding and appears to have preferred a shallow marine environment.
SPECIES *Bathyurus superbus* Raymond, *Bathyurus acutus* Raymond, *Bathyurus extans* (Hall)—Black River Group, southern Ontario and Ottawa area.
REFERENCES Wilson (1947), Whittington (1953), Ludvigsen (1978)

Raymondites Sinclair Fig. 22
This bathyurid trilobite is in many respects similar to *Bathyurus* and was probably a descendant of that genus. Compared with *Bathyurus, Raymondites* has a narrower glabella, a broader border on the cephalon, and palpebral lobes that are located further back on the cranidium. In addition, the glabella is covered by coarse tubercles. The most striking feature of the Ontario species of *Raymondites* is the presence of a pair of long curving median spines—one on the occipital ring and one on the pygidium. The name *Raymondites* is in honour of an early student of Ontario trilobites, P.E. Raymond (Fig. 16C).
SPECIES *Raymondites spiniger* (Hall)—Black River Group and Bobcaygeon and

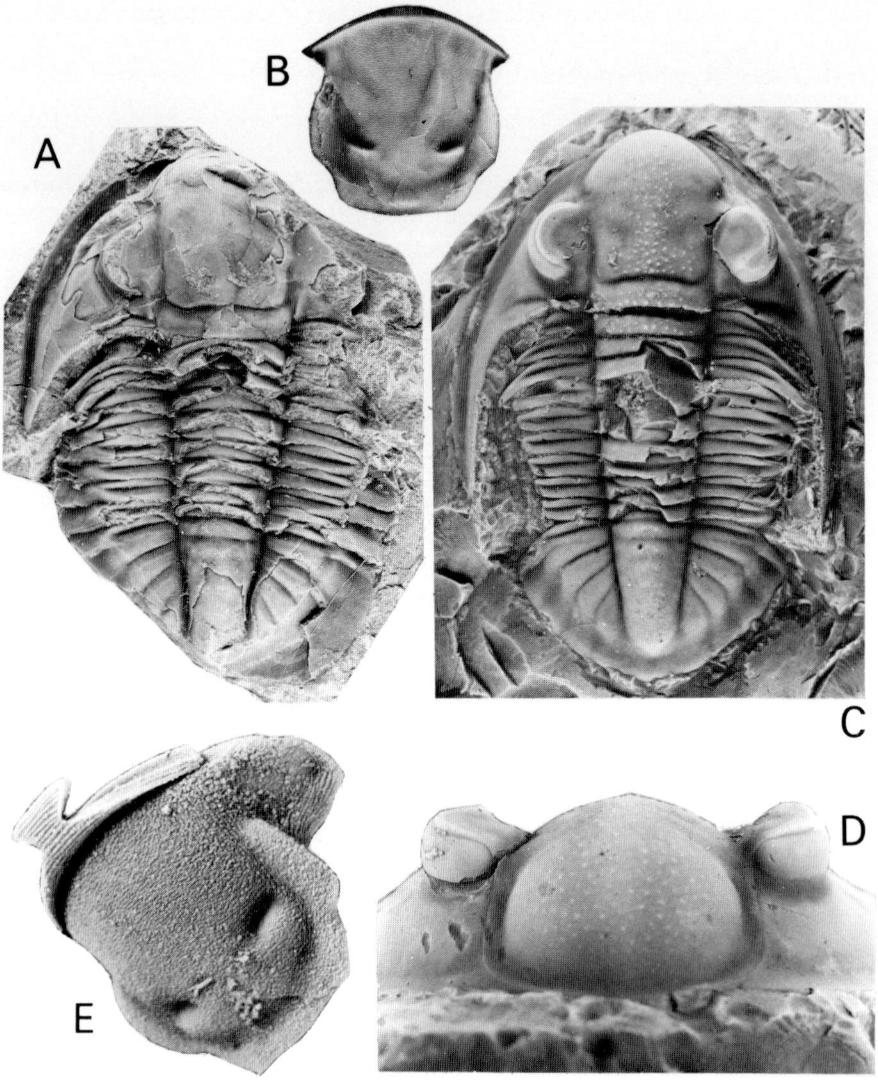

Fig. 21 ***Bathyurus***

Bathyurus superbus Raymond

A Exoskeleton, × 1, ROM 18788, Black River Gp., Ottawa.

B Hypostome, × 1.7, ROM 32374, Black River Gp., Ottawa.

Bathyurus extans (Hall)

C,D Exoskeleton, × 2 and × 3.5, USNM 306, Lowville Fm., Great Bend, New York.

Bathyurus sp.

E Hypostome and attached rostral plate, × 8.2, GSC 40431, Esbataottine Fm., South Nahanni River area, District of Mackenzie.

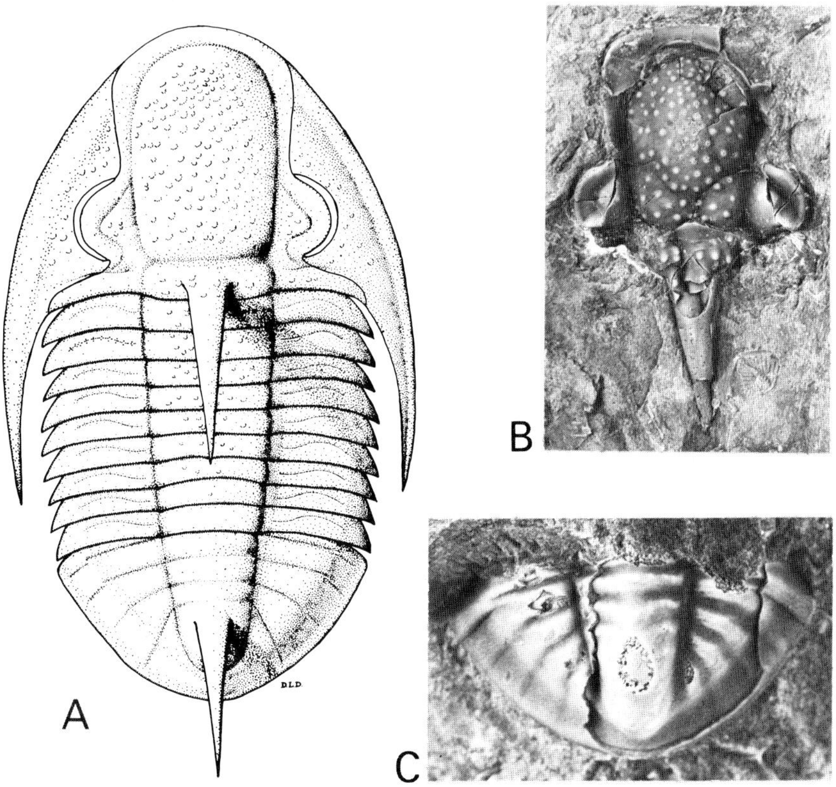

Fig. 22 ***Raymondites***
Raymondites spiniger (Hall)
A Reconstruction, × 1.7.
B Cranidium, × 1.5, GSC 4318, Bobcaygeon Fm., Kirkfield Quarry, Kirkfield.
C Pygidium, × 3.5, ROM 35353, Verulam Fm., Moira River, Belleville.

Verulam Formations, southern Ontario. Other species are *Raymondites bandifer* Sinclair and *R. trispinosus* (Wilson).
REFERENCES Sinclair (1944), Wilson (1947), Whittington (1953)

***Failleana* Chatterton and Ludvigsen Fig. 23**
Failleana is a large, smooth and slightly inflated illaenid trilobite that is occasionally found in the limestones of the upper Black River and Trenton Groups of southern Ontario. It is virtually devoid of furrows except for shallow axial furrows on the cephalon. It is thus sometimes difficult to recognize isolated cranidia and pygidia of this genus as belonging to a trilobite.
SPECIES *Failleana indeterminata* (Walcott)—upper Black River and Trenton Groups, southern Ontario and Ottawa area.
REFERENCES Raymond (1916), Wilson (1947), Chatterton and Ludvigsen (1976)

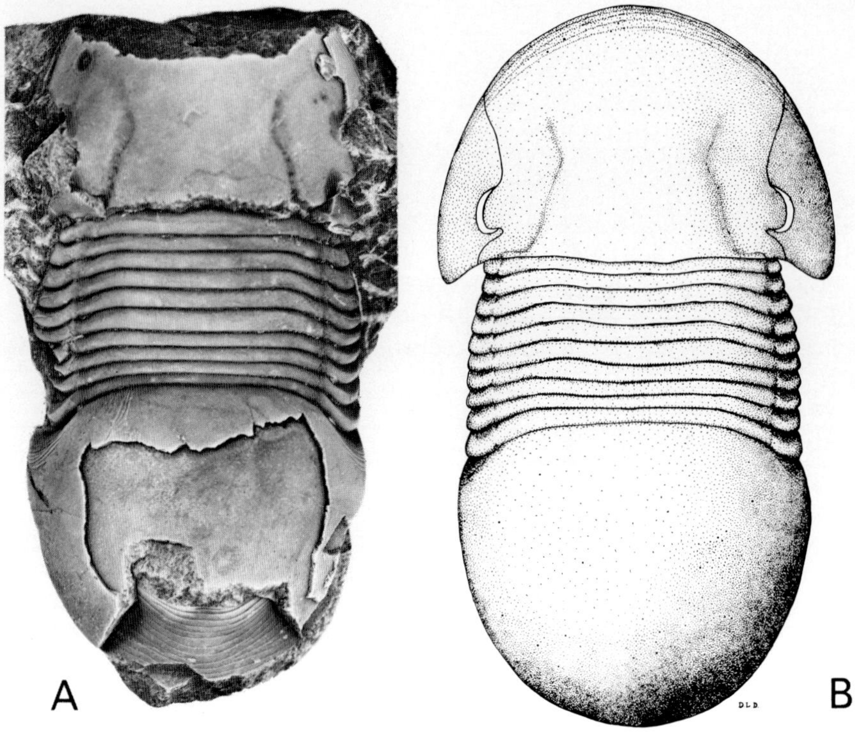

Fig. 23 ***Failleana***
Failleana indeterminata (Walcott)
A Exoskeleton, × 1, GSC 331, Cobourg Fm., Ottawa.
B Reconstruction, × 1.

Calyptaulax Cooper Fig. 24A,B

Calyptaulax (or *Calliops* as it has often been called) is one of the more commonly encountered trilobites in the Ordovician limestones of Ontario. This dalmanitid trilobite is characterized by a bulbous glabella with three pairs of straight lateral furrows, large crescent-shaped palpebral lobes with deep palpebral furrows, and blunt genal angles. The glabella is covered by large granules. The pygidium is triangular with a well-defined axis and is strongly ribbed. The aggregate eyes of *Calyptaulax* are large and strongly curved in a horizontal plane. Each eye is comprised of about 130 small domelike lenses which are distinctly separated from each other and arranged in vertical rows. Each lens is covered by a discrete cornea and is oriented slightly differently than its neighbour. This permits the accumulation of visual information from a wide horizontal field—in fact, from all directions except immediately in front of and behind the trilobite. Levi-Setti (1976) gave a concise account of trilobite eyes and trilobite vision, including discussion of genera very similar to *Calyptaulax*.

SPECIES *Calyptaulax callicephalus* (Hall)—at almost every locality of the upper Black River and Trenton Groups, southern Ontario and Ottawa area.

REFERENCES Delo (1940), Cooper (1953), Shaw (1968, 1974), Chatterton and Ludvigsen (1976)

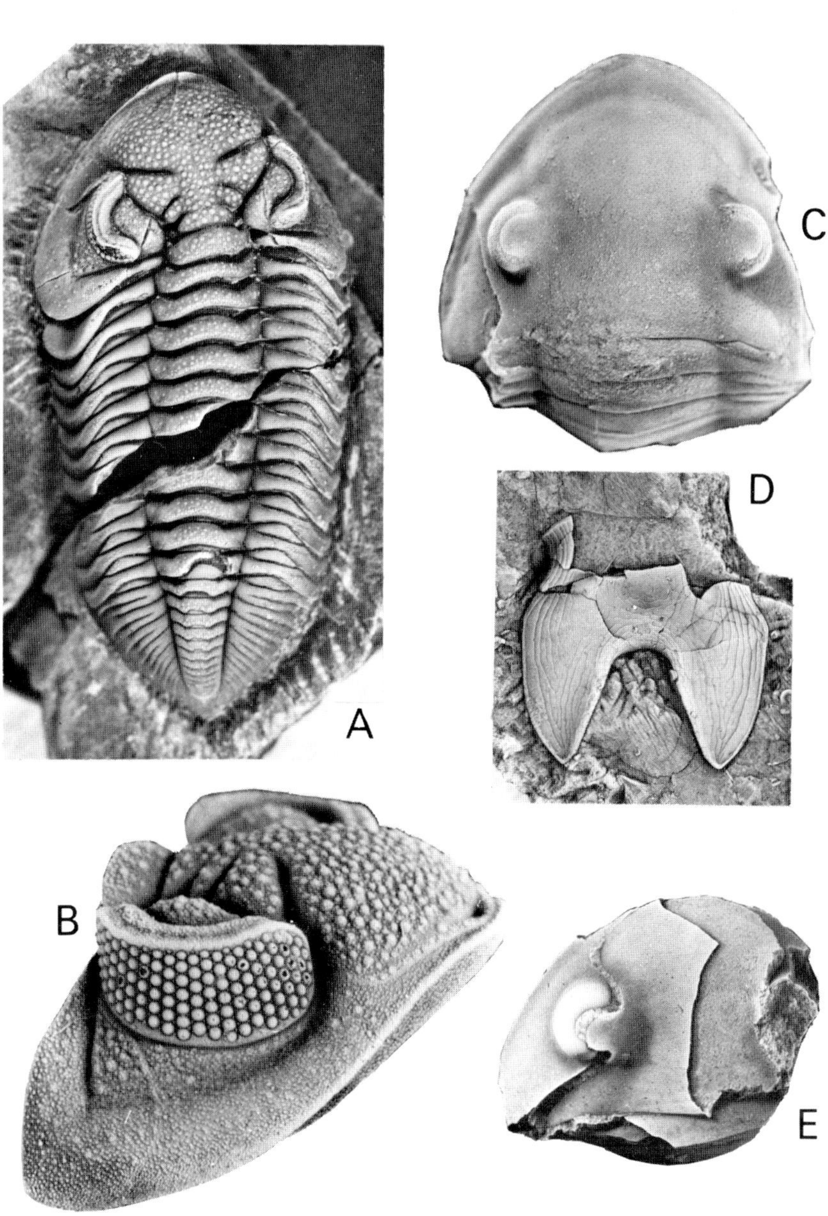

Fig. 24 ***Calyptaulax Isotelus Anataphrus***

Calyptaulax callicephalus (Hall)

A Exoskeleton, × 3, ROM 35167, Verulam Fm., Gamebridge.

Calyptaulax callirachis (Cooper)

B Silicified cephalon, × 9.5, GSC 40438, Esbataottine Fm., South Nahanni River area, District of Mackenzie.

Isotelus gigas DeKay

C Enrolled specimen, × 2.6, ROM 35367, Verulam Fm., Lakefield Quarry, Lakefield.

D Hypostome, × 1.7, ROM 35368, Verulam Fm., Lakefield Quarry, Lakefield.

Anataphrus sinclairi (Wilson)

E Cephalon, × 3.5, GSC 13252, Black River Gp., Ottawa.

***Anataphrus* Whittington Fig. 24E**

Anataphrus means "without furrows" and this asaphid trilobite is aptly named. The absence of axial, glabellar, and border furrows lends a globose aspect to this trilobite and distinguishes it from the related genus *Isotelus*.

SPECIES *Anataphrus sinclairi* (Wilson)—Black River Group, Ottawa area.

REFERENCES Wilson (1947), Whittington (1954)

***Isotelus* DeKay Figs. 17, 24C,D, 25**

This asaphid trilobite is, without question, the most frequently encountered trilobite in Ordovician rocks of Ontario. Species of *Isotelus* occur in a wide variety of rock types, including limestones, shales, and siltstones, and often attain impressive dimensions. The specimen illustrated in Figure 25 measures almost 30 cm in length and larger specimens have been found. *Isotelus* has a smooth triangular cephalon with flattened borders and shallow axial furrows. The eyes are prominent and slightly elevated. The axial furrows are distinct on the thorax, but very faint on the triangular pygidium. The distinctive hypostome is rectangular in outline and is deeply notched posteriorly. In front of the hypostome occurs a broad and flat doublure that is covered with fine ledgelike lines that parallel the margin of the cephalon.

SPECIES A large number of species of *Isotelus* have been reported from Ontario; among them *Isotelus gigas* DeKay, *Isotelus platycephalus* (Stokes), *Isotelus maximus* Locke, *Isotelus iowensis* (Owen), *Isotelus ottawaensis* Wilson, and *Isotelus latus* Raymond.

REFERENCES Raymond (1921), Wilson (1947), Darby and Stumm (1965), Chatterton and Ludvigsen (1976)

***Ceraurus* Green Fig. 26A–D**

It is a rare Ordovician limestone outcrop in Ontario that does not yield this distinctive and often large cheirurid trilobite. *Ceraurus* has an inflated and forwardly expanding glabella with three pairs of deep lateral furrows. The eyes are generally located far out on the cheek and the genal angles are extended into a pair of long and curving genal spines. The cephalon is covered by tubercles of varying sizes. The semicircular pygidium is enhanced in size by a pair of long curving first spines, behind which may occur another two pairs of small spines. The hypostome is oval in outline and has an inflated central portion and a rolled border.

SPECIES *Ceraurus pleurexanthemus* Green, *Ceraurus dentatus* Raymond and Barton, *Ceraurus bispinosus* Raymond and Barton—upper Black River and Trenton Groups, southern Ontario and Ottawa area.

REFERENCES Raymond and Barton (1913), Raymond (1921), Wilson (1947), Cooper (1953), Evitt (1953), Lane (1971)

***Eomonorachus* Delo Fig. 26E**

Eomonorachus is similar to *Calyptaulax*, but differs from that dalmanitid trilobite by possessing smaller palpebral lobes which are located closer to the axial furrows, long genal spines, and sparser granulate ornament on the diamond-shaped glabella. In addition, the triangular pygidium has an axis that extends

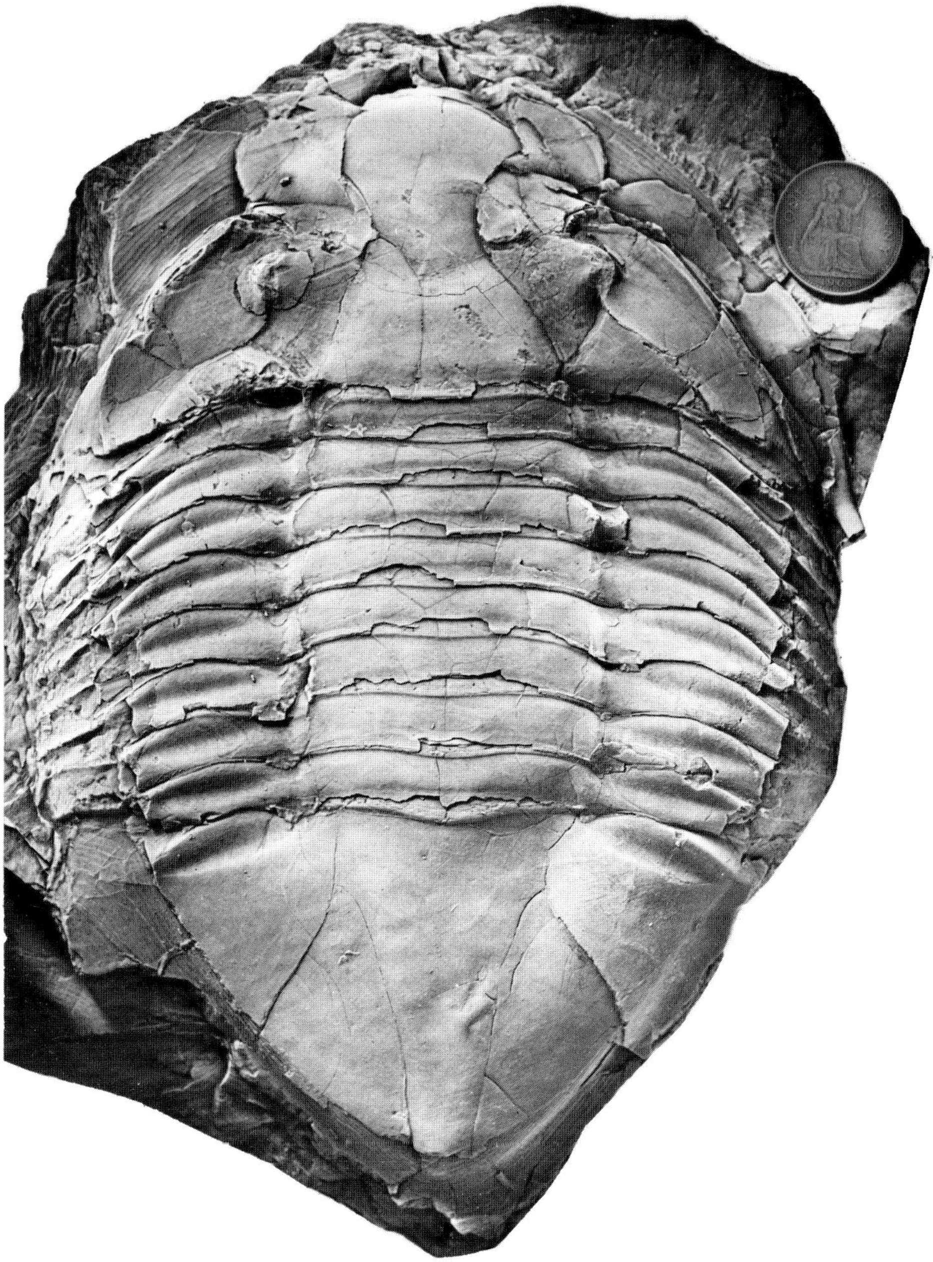

Fig. 25 ***Isotelus***
Isotelus maximus Locke
Large specimen, × .65, ROM 189, Georgian Bay Fm., Don Valley Brickyards, Toronto.

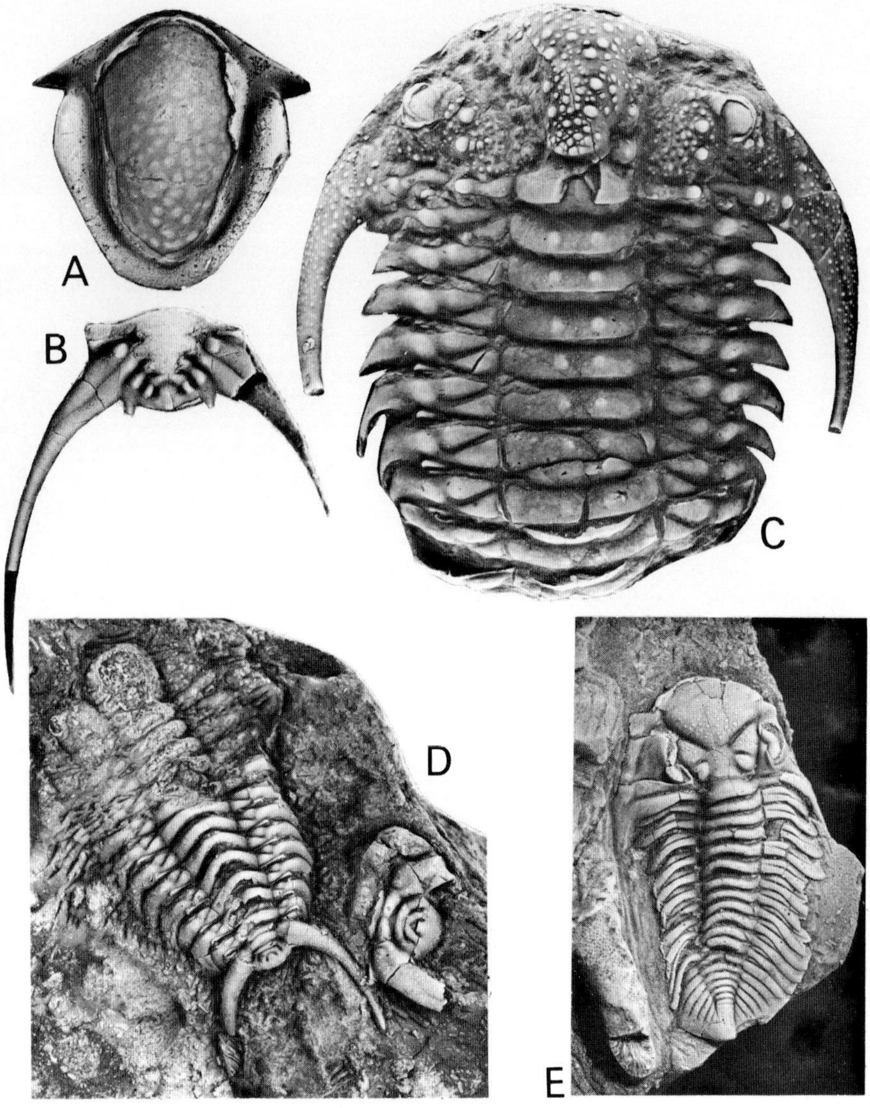

Fig. 26 ***Ceraurus Eomonorachus***

Ceraurus sp.

A Hypostome, × 2.3, ROM 35363, Verulam Fm., Moira River, Belleville.

B Pygidium, × 1.3, ROM 35364, Bobcaygeon Fm., Kirkfield Quarry, Kirkfield.

Ceraurus dentatus Raymond and Barton

C Exoskeleton, × 1.9, GSC 1769, Verulam Fm., Belleville.

Ceraurus pleurexanthemus Green

D Crushed specimen and pygidium, × 1.7, ROM 35366, Verulam Fm., Moira River, Belleville.

Eomonorachus intermedius (Walcott)

E Exoskeleton, × 1.7, GSC 57761, Trenton Gp., Goat Island.

nearly to the tip of the pygidium and the ribs on the flanks of the pygidium are unfurrowed. *Eomonorachus* is a rather uncommon trilobite in Ontario.
SPECIES *Eomonorachus intermedius* (Walcott)—Trenton Group, Goat Island.
REFERENCE Delo (1940)

Encrinuroides Reed Fig. 27
Encrinuroides is one of the more common trilobites in the Bobcaygeon and Verulam Formations of southern Ontario. It is readily distinguished by its nearly parallel-sided glabella which is covered by large blunt tubercles. These tubercles largely obscure three short lateral glabellar furrows. The front portion of the glabella has a short longitudinal furrow. The Ontario species of *Encrinuroides* bears long curving genal spines and a single median axial spine on the seventh and ninth thoracic segment and paired spines on the pleural lobe of the eighth. The pygidium is triangular and has numerous axial rings and 10 ribs that curve downward and backward and terminate as short blunt spines.
SPECIES *Encrinuroides vigilans* (Hall)—Bobcaygeon and Verulam Formations, southern Ontario. This species masqueraded under a number of names, including *Encrinurus trentonensis* Walcott and *Encrinurus cybeleformis* Raymond.
REFERENCE Evitt and Tripp (1977)

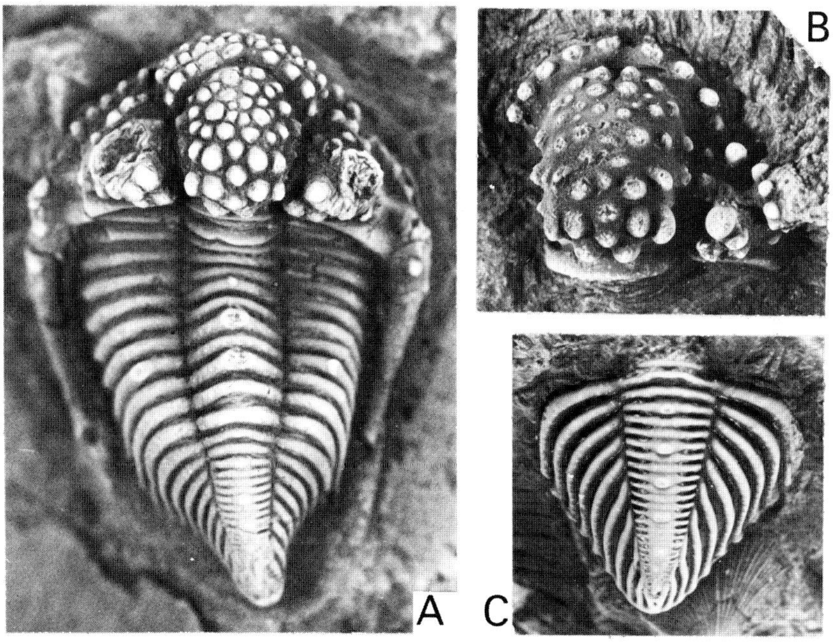

Fig. 27 ***Encrinuroides***
Encrinuroides vigilans (Hall)
A Exoskeleton, × 5.2, AMNH 36070, Trenton Gp., Middleville, New York.
B Cranidium, × 3.9, ROM 35357, Verulam Fm., Moira River, Belleville.
C Pygidium, × 3.9, ROM 35358, Verulam Fm., Moira River, Belleville.

***Achatella* Delo Fig. 28A,B**

Achatella is a rather uncommon dalmanitid trilobite in the Verulam Formation. The cephalon is lowly vaulted, crescentic in outline, and has a pair of long slim genal spines. The sparsely tuberculate glabella expands markedly forward and has three pairs of straight lateral glabellar furrows. The palpebral lobes are small and nearly vertical and the visual surface of the eye is high and triangular.

SPECIES *Achatella achates* (Billings)—Verulam and Cobourg Formations, southern Ontario and Ottawa area.

REFERENCES Delo (1940), Sinclair (1944), Wilson (1947)

***Bumastoides* Whittington Fig. 28C,D**

This illaenid trilobite is generally represented by discrete cranidia and pygidia; both are distressingly nondescript in their oval outline, inflation, and lack of ornamentation or distinct furrows. Internal moulds of the cranidium, however, show shallow axial furrows which terminate in crescent-shaped depressions ("lunettes"). The complete trilobite is elliptical in outline, smooth, and lacks projections of any kind. *Bumastoides* was probably a burrowing trilobite.

SPECIES *Bumastoides billingsi* (Raymond and Narraway) and *Bumastoides milleri* (Billings)—upper Black River and Trenton Groups, southern Ontario and Ottawa area.

REFERENCES Raymond and Narraway (1906), Wilson (1947), Whittington (1954), Chatterton and Ludvigsen (1976)

"*Calyptaulax*" Fig. 28E,F

Species of this genus have hitherto been assigned to *Calyptaulax*, but well-preserved specimens from the Verulam Formation suggest that they belong to a new unnamed genus. "*Calyptaulax*" differs from *Calyptaulax* in having a larger anterior portion of the glabella, eyes situated further away from the axial furrows, long tapering genal spines, and a narrower and higher pygidium.

SPECIES "*Calyptaulax*" *lincolnensis* (Branson)—Bobcaygeon and Verulam Formations, southern Ontario. Other species, such as *Calliops alatus* Okulitch, *Calyptaulax ottawanus* Okulitch, and *Calyptaulax calderi* Wilson probably belong in "*Calyptaulax*".

REFERENCES Delo (1940), Wilson (1947)

***Flexicalymene* Shirley Figs. 29, 45A,B**

Flexicalymene is one of the more abundant trilobites in Ordovician strata of Ontario, probably second only to *Isotelus*. The name refers to the common occurrence of this trilobite as enrolled specimens, but such preservation is not unique to this trilobite. *Flexicalymene* is characterized by a moderately convex glabella of parabolic outline and a broad trough-shaped preglabellar furrow. Three pairs of lateral glabellar lobes decrease in size towards the front. The pygidium has a broad axis and lacks a border. The elongate hypostome bears a pair of pointed posterior projections. The entire exoskeleton is finely granulate. A single species occurs in Silurian rocks of southern Ontario.

SPECIES *Flexicalymene senaria* (Conrad)—Trenton Group, *Flexicalymene granulosa* (Foerste)—Georgian Bay Formation, *Flexicalymene celebra* (Raymond)—Amabel Formation, southern Ontario.

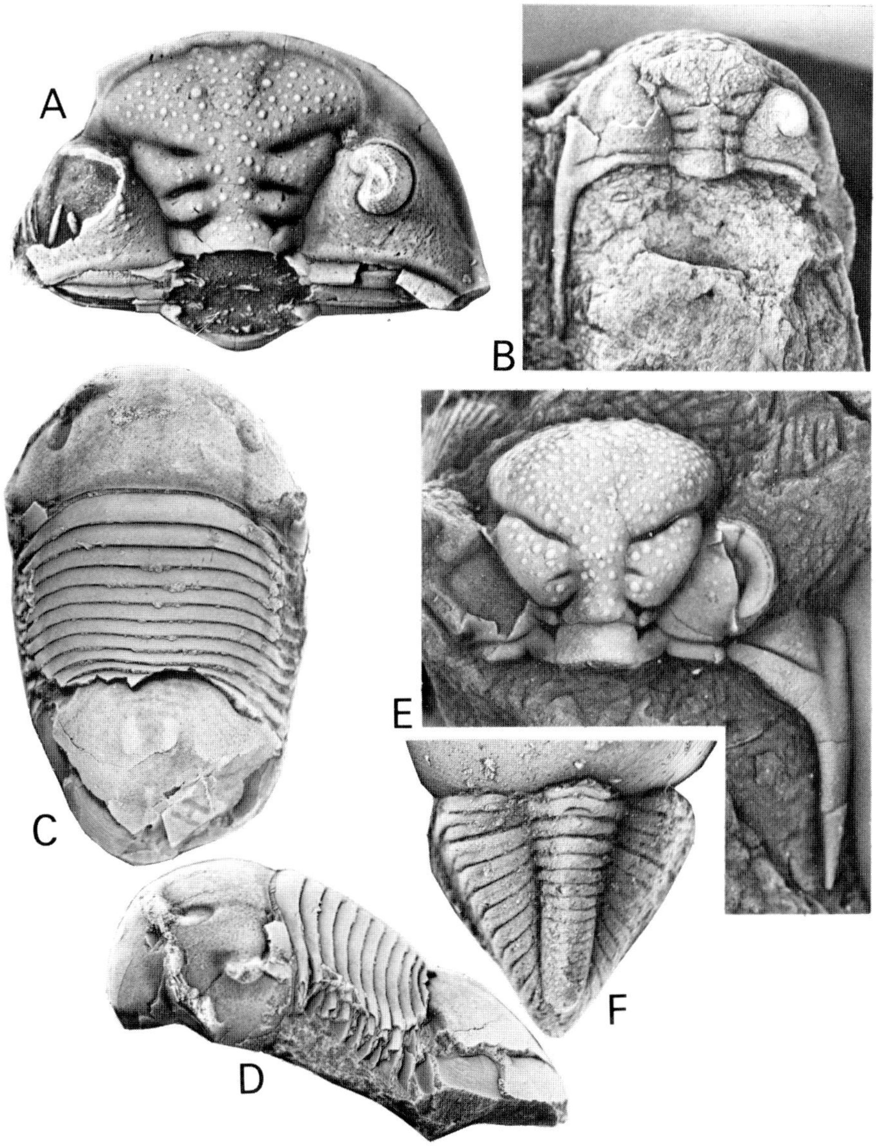

Fig. 28 ***Achatella Bumastoides "Calyptaulax"***
Achatella achates (Billings)
A Cephalon, × 4.3, GSC 13275, Verulam Fm., Lakefield Quarry, Lakefield.
B Cephalon, × 3, ROM 35372, Verulam Fm., Lakefield Quarry, Lakefield.
Bumastoides billingsi (Raymond and Narraway)
C,D Exoskeleton, two views, × 1.2, ROM 18792, Trenton Gp., Hull, Quebec.
"Calyptaulax" lincolnensis (Branson)
E Cranidium, × 4.3, ROM 35373, Verulam Fm., Lakefield Quarry, Lakefield.
F Pygidium, × 4.9, ROM 35374, Verulam Fm., Moira River, Belleville.

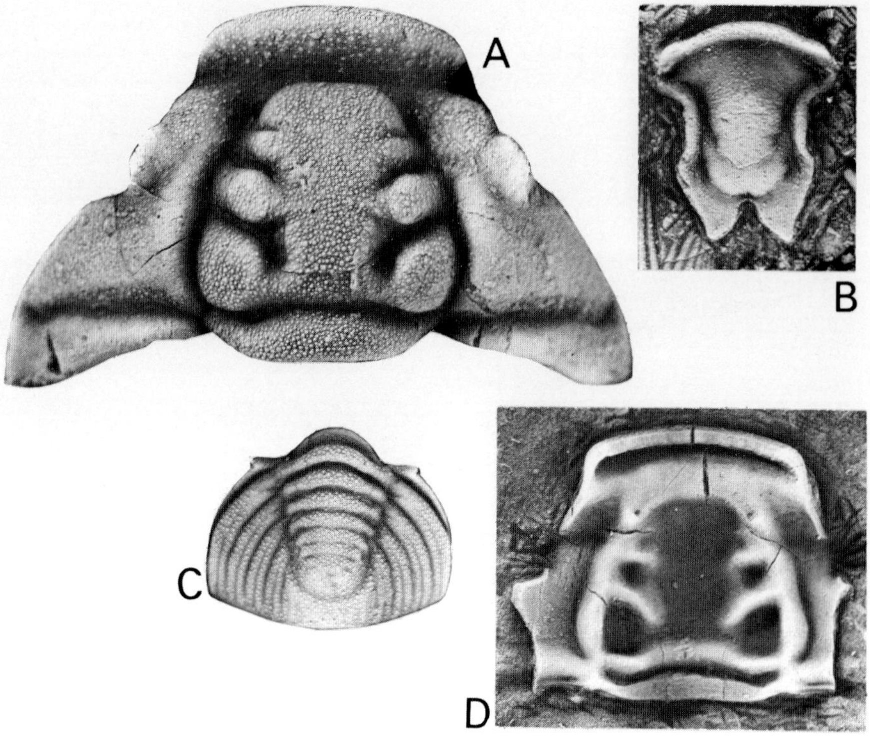

Fig. 29 ***Flexicalymene***
Flexicalymene senaria (Conrad)
A Cranidium, × 4.7, ROM 35359, Verulam Fm., Lakefield Quarry, Lakefield.
B Hypostome, × 6, ROM 35360, Verulam Fm., Lakefield Quarry, Lakefield.
C Pygidium, × 5.2, ROM 35361, Verulam Fm., Lakefield Quarry, Lakefield.
D Cranidium, ventral view, × 4.3, ROM 35362, Verulam Fm., Lakefield Quarry, Lakefield.

REFERENCES Whittington (1941, 1971), Evitt and Whittington (1953), Stumm and Kauffman (1958), Ross (1967), Siveter (1976)

***Paraharpes* Whittington Fig. 30A**
Paraharpes is one of the most beautiful and distinctive of the Ontario trilobites and also one of the rarest. The convex glabella is parabolic in outline and stands well above the remaining cephalon. The cheeks of this harpid trilobite are slightly inflated and carry a fine reticulate ornament. This portion of the cephalon is surrounded by a yoke-shaped brim which is broad, flat, and finely pitted. The eyes are small and evident as low rounded tubercles on the front part of the cheeks.
SPECIES *Paraharpes ottawaensis* (Billings)—Cobourg Formation, Ottawa area.
REFERENCES Wilson (1947), Whittington (1950)

***Dolichoharpes* Whittington Fig. 30B**
Dolichoharpes is somewhat similar to *Paraharpes*, but it has a much narrower cephalon, steeper cephalic margins, and a narrower and upturned brim. The un-

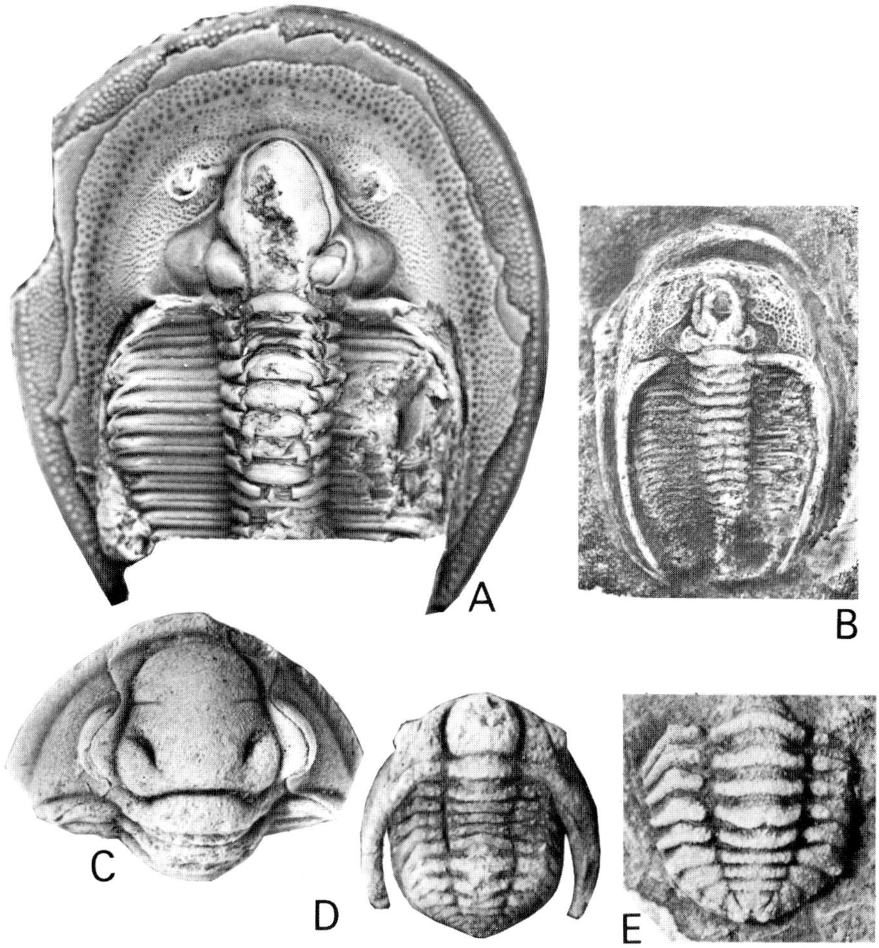

Fig. 30 ***Paraharpes Dolichoharpes Cyphoproetus Dimeropyge***
Paraharpes ottawaensis (Billings)
A Exoskeleton, × 2.3, ROM 329, Cobourg Fm. (?), Ottawa.
Dolichoharpes dentoni (Billings)
B Exoskeleton, × 2.3, GSC 1767, Trenton Gp., Ottawa.
Cyphoproetus wilsonae (Sinclair)
C Enrolled specimen, × 6, GSC 13255, lower Trenton Gp., Eganville.
Dimeropyge gibbus Sinclair
D Exoskeleton, × 8.6, GSC 7828, Verulam Fm., Peterborough.
E Thorax and pygidium, × 8.6, GSC 6701, Verulam Fm., Lakefield.

dersurface of the brim is broad and flat. Such a surface acted in a similar manner to a snowshoe and probably served the purpose of preventing the trilobite from sinking into the soft mud.

SPECIES *Dolichoharpes dentoni* (Billings)—upper Trenton Group, southern Ontario and Ottawa area.

REFERENCES Whittington (1949), Chatterton and Ludvigsen (1976)

Cyphoproetus Kegel Fig. 30C

This tiny trilobite has a moderately inflated glabella that narrows towards the front and has two pairs of lateral furrows—the first deep and oblique, the second shallow and transverse. The palpebral lobes are crescentic and moderately small. The cephalon has a narrow convex border.

SPECIES *Cyphoproetus wilsonae* (Sinclair)—lower Trenton Group, southern Ontario.

REFERENCE Owens (1973)

Dimeropyge Öpik Fig. 30D,E

Dimeropyge is probably more common than the half dozen or so known Ontario specimens indicate, but it is so tiny that it is probably overlooked by collectors. The complete specimen in Figure 30D is less than 4 mm in total length! *Dimeropyge* is characterized by a rounded and inflated glabella, widely set eyes, and long genal spines. The semicircular pygidium has three segments and a steep posterior margin which is rimmed by small spines. The entire exoskeleton is studded with tiny granules and spines.

SPECIES *Dimeropyge gibbus* Sinclair—Verulam Formation, southern Ontario.

REFERENCES Sinclair (1946), Whittington and Evitt (1954), Chatterton and Ludvigsen (1976)

Hypodicranotus Whittington Fig. 31A,B

Hypodicranotus is the only remopleuridid trilobite found in Ontario. It has a broad smooth glabella that occupies the total area between the long striplike eyes and long genal spines. The axis on the cephalon and thorax is very broad. The extremely long and deeply notched hypostome extends nearly to the posterior tip of the trilobite and this is the most characteristic element of the genus.

SPECIES *Hypodicranotus striatulus* (Walcott)—Verulam Formation, southern Ontario.

REFERENCE Whittington (1952)

Cybeloides Slocum Fig. 31C

Cybeloides has an inflated glabella that is partitioned into three longitudinal lobes by a pair of furrows, each of which comprises four deep pits. The eyes are stalked and the pygidium is high and narrowly triangular. The genus is very rare in Ordovician rocks of Ontario, although moderately common in other areas.

SPECIES *Cybeloides ella* (Raymond and Narraway)—Black River Group, Ottawa area and *Cybeloides plana* Sinclair—Verulam Formation, southern Ontario.

REFERENCES Wilson (1947), Cooper (1953), Chatterton and Ludvigsen (1976), Evitt and Tripp (1977)

Cryptolithus Green Fig. 31D,E

This trinucleid trilobite is easily recognized by its smooth inverted drop-shaped glabella, short occipital spine, and broad pitted fringe on the cephalon. The pits on the fringe are distinctly arranged in radiating and concentric rows. *Cryptolithus* lacks any trace of eyes and must be assumed to have been blind. It

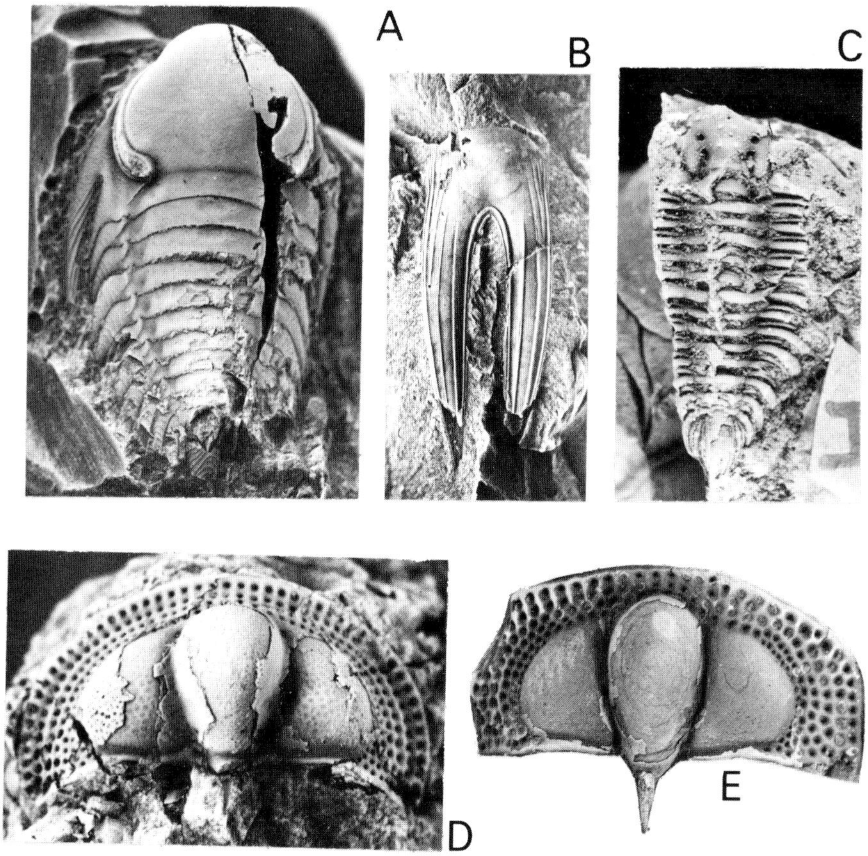

Fig. 31 ***Hypodicranotus Cybeloides Cryptolithus***
Hypodicranotus striatulus (Walcott)
A Exoskeleton, × 2.3, MCZ 1616, Middle Trenton Gp., Trenton Falls, New York.
B Hypostome, × 2.3, MCZ 1617, Middle Trenton Gp., Trenton Falls, New York.
Cybeloides ella (Raymond and Narraway)
C Exoskeleton, × 4.1, ROM 18696, Black River Gp., Ottawa.
Cryptolithus lorettensis Foerste
D Cephalon, × 4.7, ROM 746 T, Sherman Falls Limestone, Montmorency Falls, Quebec.
Cryptolithus bellulus (Ulrich)
E Cephalon, × 3, GSC 18669, upper Whitby Fm., Meaford.

is found at two distinct stratigraphic levels in southern Ontario—in the Verulam Formation and in the upper Whitby to Georgian Bay Formations.

SPECIES *Cryptolithus lorettensis* Foerste—Verulam Formation and *Cryptolithus bellulus* (Ulrich)—upper Whitby and Georgian Bay Formations, southern Ontario and Ottawa area.

REFERENCES Whittington (1941, 1959, 1968), Hughes *et al.* (1975), Campbell (1975)

***Eobronteus* Reed Fig. 32A,B**

Eobronteus is an uncommon element of the Ordovician faunas of Ontario. It is a

moderately large and flat scutelluid trilobite. The cephalon is crescentic in outline and has a broad flat border. The glabella is only slightly convex and expands markedly towards the front. The eyes are located near the posterior edge of the cephalon. The pygidium is large and semicircular and possesses six paired ribs arranged in a fan-shape about a short blunt axis. The exoskeleton bears an ornament of fine raised lines which are transversely oriented.
SPECIES *Eobronteus lunatus* (Billings) and *Eobronteus sculptus* Sinclair—Verulam and Cobourg Formations, southern Ontario and Ottawa area.
REFERENCES Sinclair (1949), Šnajdr (1960)

***Ceraurinella* Cooper Fig. 32c**
This cheirurid trilobite is rather similar to *Ceraurinus*, but has a more inflated glabella, lateral glabellar furrows that curve, and a cephalic border that is convex. The pygidium is characteristic in that it has first spines that are directed obliquely upward and backward. *Ceraurinella* is a common trilobite in Ordovician rocks in areas such as Virginia and the Northwest Territories, but is surprisingly rare in Ontario.
SPECIES *Ceraurinella trentonensis* (Barton)—Trenton Group, Goat Island.
REFERENCES Cooper (1953), Whittington and Evitt (1954), Chatterton and Ludvigsen (1976)

***Sphaerocoryphe* Angelin Fig. 32d**
Complete specimens of this small cheirurid trilobite are exceedingly rare, but the distinctive spherical front portion of the glabella is not uncommon in the limestones and shales of the Verulam Formation. The pygidium consists largely of a pair of curved pincer-shaped spines.
SPECIES *Sphaerocoryphe robustus* Walcott—Verulam Formation, southern Ontario.
REFERENCES Shaw (1968), Lane (1971)

***Hemiarges* Gürich Fig. 33**
The size of this small lichid trilobite makes it difficult to identify, but the distinctive tripartite glabella with its inflated lateral lobes can be seen readily with the aid of a hand-lens. The pygidium has a short axis and four pairs of unequal spines. The hypostome is transversely rectangular. *Hemiarges* is covered by tubercles of varying sizes. The genus is most common in the Verulam Formation of southern Ontario.
SPECIES *Hemiarges paulianus* (Clarke)—upper Trenton Group, southern Ontario and Manitoulin Island.
REFERENCES Foerste (1920), Tripp (1957), Chatterton and Ludvigsen (1976)

***Amphilichas* Raymond Fig. 34**
Amphilichas has a semicircular cranidium that is dominated by an inflated glabella. A pair of longitudinal furrows divides the glabella into three lobes—the median lobe is wider and more inflated than the lateral lobes. The hypostome is oval in outline and indented posteromedially. The pygidium has a goblet-shaped axis which is flanked by three pairs of broad backwardly directed spines. *Amphilichas* (or *Acrolichas* as it has commonly been called) is a rather uncom-

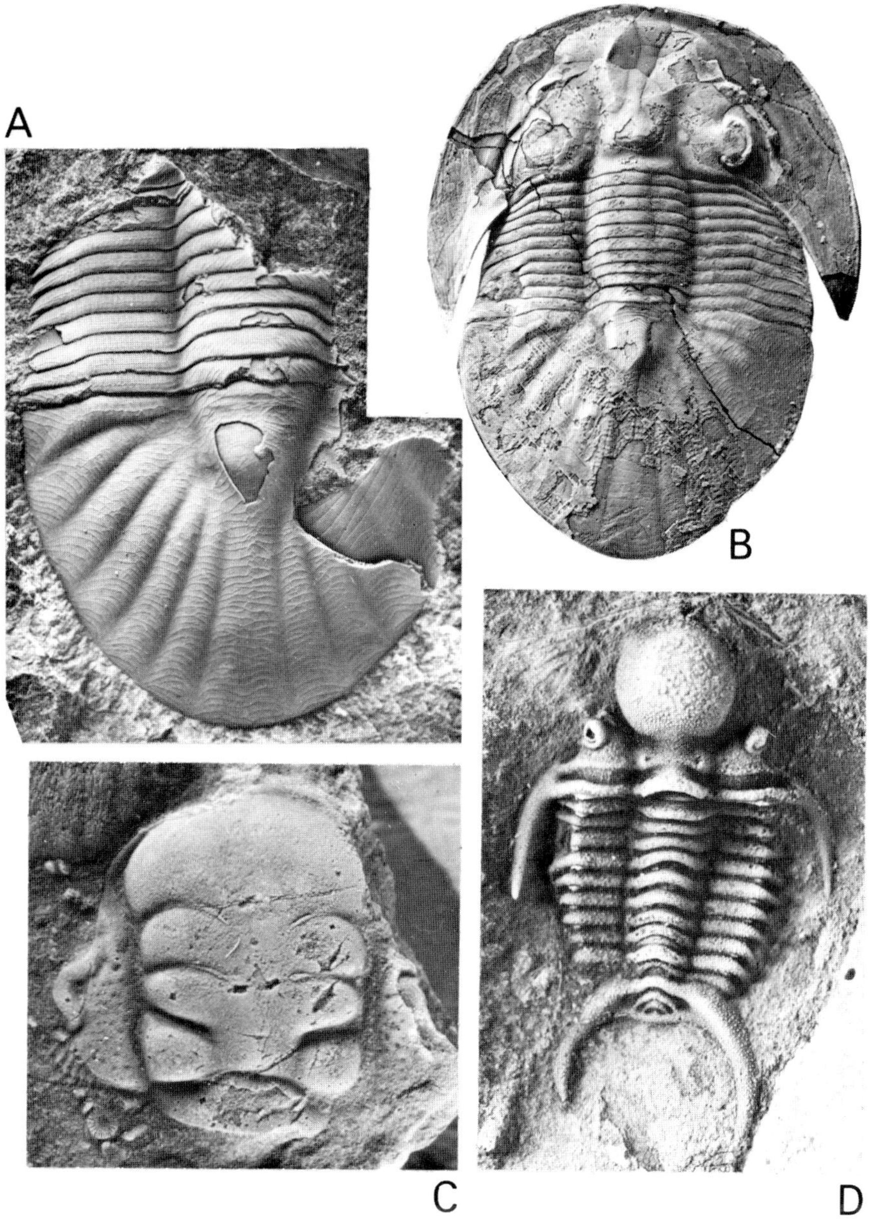

Fig. 32 ***Eobronteus Ceraurinella Sphaerocoryphe***
Eobronteus sculptus Sinclair
A Exoskeleton, × 1.8, ROM 18734a, Trenton Gp., St. Louis Dam, Ottawa.
Eobronteus lunatus (Billings)
B Exoskeleton, × 1.6, GSC 1781, Trenton Gp., Ottawa.
Ceraurinella trentonensis (Barton)
C Cranidium, × 4.5, GSC 5836, Trenton Gp., Goat Island.
Sphaerocoryphe robustus Walcott
D Exoskeleton, × 5.2, MCZ 7567, Middle Trenton Gp., Trenton Falls, New York.

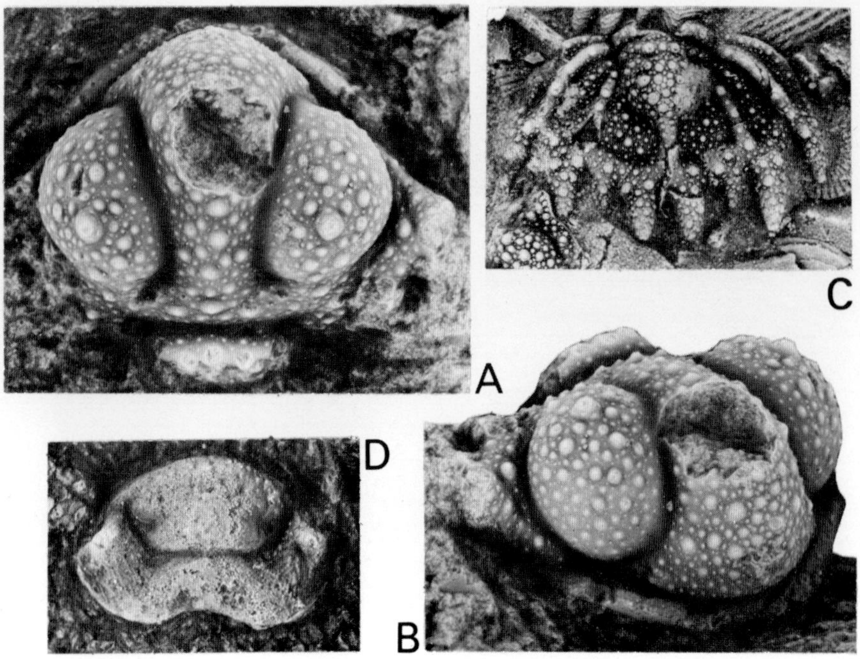

Fig. 33 ***Hemiarges***
Hemiarges paulianus (Clarke)
A,B Cranidium, × 6.9, ROM 35355, Verulam Fm., Moira River, Belleville.
C Pygidium, × 6.9, USNM 43436a, Prosser Limestone, St. Paul, Minnesota.
D Hypostome, × 5, ROM 35356, Verulam Fm., Lakefield Quarry, Lakefield.

mon trilobite in the Trenton Group of southern Ontario; certainly less common than *Hemiarges*, the other lichid trilobite.
SPECIES *Amphilichas ottawaensis* (Foerste) and *Amphilichas narrawayi* (Foerste) —Trenton Group, southern Ontario and Ottawa area.
REFERENCES Tripp (1957), Shaw (1968), Chatterton and Ludvigsen (1976)

Ceraurinus Barton Fig. 35

This cheirurid trilobite is relatively common in the Cobourg Formation and also occurs in Upper Ordovician rocks on Manitoulin Island and in the Hudson Bay area. *Ceraurinus* is characterized by a rectangular, smooth, weakly inflated glabella with three pairs of deep and straight lateral furrows, flat cephalic borders, and short, stout genal spines. The pygidium bears three pairs of spines: the first pair is long, slightly curving, and blunt; the remaining two pairs are pointed. In addition, the first spine pair possesses a narrow median rib.
SPECIES *Ceraurinus marginatus* Barton—Cobourg Formation, southern Ontario. *Ceraurinus icarus* (Billings)—Kagawong Formation, Manitoulin Island and Upper Ordovician rocks in the Hudson Bay area.
REFERENCE Ludvigsen (1977)

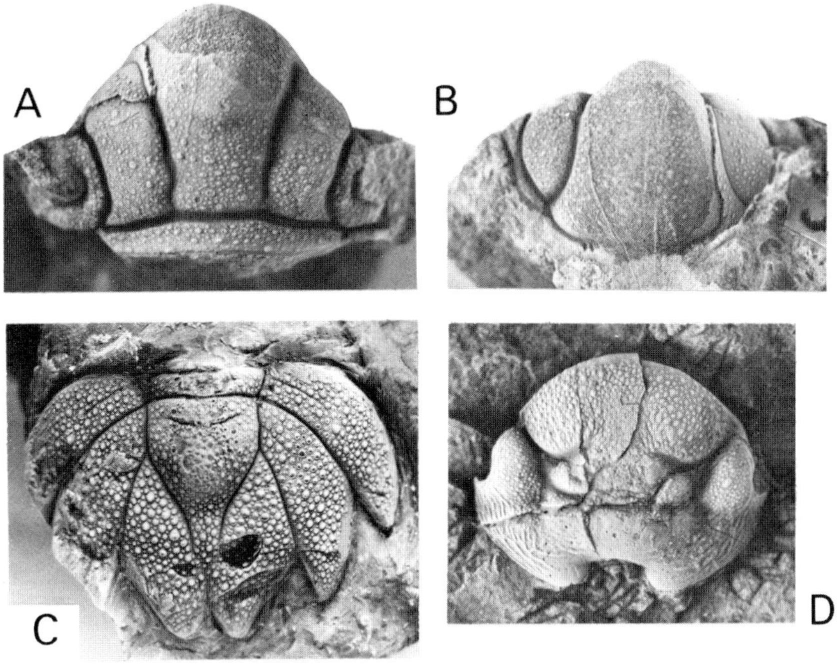

Fig. 34 ***Amphilichas***
Amphilichas ottawaensis (Foerste)
A,B Cranidium, × 1.7, ROM 18791b, Trenton Gp., Hull, Quebec.
C Pygidium, × 2.1, ROM 18791a, Trenton Gp., Hull, Quebec.
D Hypostome, × 3.4, ROM 35370, Verulam Fm., Lakefield Quarry, Lakefield.

***Pseudogygites* Kobayashi Figs. 10, 14, 36, 37**

The brown shales of the lower Whitby Formation exposed at the shores of Georgian Bay near Craigleith and on Lake Ontario near Whitby contain vast numbers of this asaphid trilobite. *Pseudogygites* has a glabella that expands markedly forward, eyes located close to the axial furrows, a distinct and wide flattened border on the cephalon, and spikelike genal spines. The pygidium is subtriangular in outline and strongly ribbed. The general aspect of the exoskeleton of *Pseudogygites* indicates a close relationship to *Isotelus*. Unlike other asaphid trilobites, the species of *Pseudogygites* that occurs in southern Ontario does not possess a median suture crossing the doublure (see Fig. 10).
SPECIES *Pseudogygites latimarginatus* (Hall)—upper Cobourg and lower Whitby Formations, southern Ontario and Ottawa area.
REFERENCE Ludvigsen (in press)

***Triarthrus* Green Figs. 38, 39**

This small trilobite is the only representative of the olenids in Ontario. *Triarthrus* occurs in great numbers in the brown shales of the lower Whitby Formation, virtually covering entire bedding surfaces in places. The small size and oval outline of the exoskeleton and the wide axis and square glabella make this

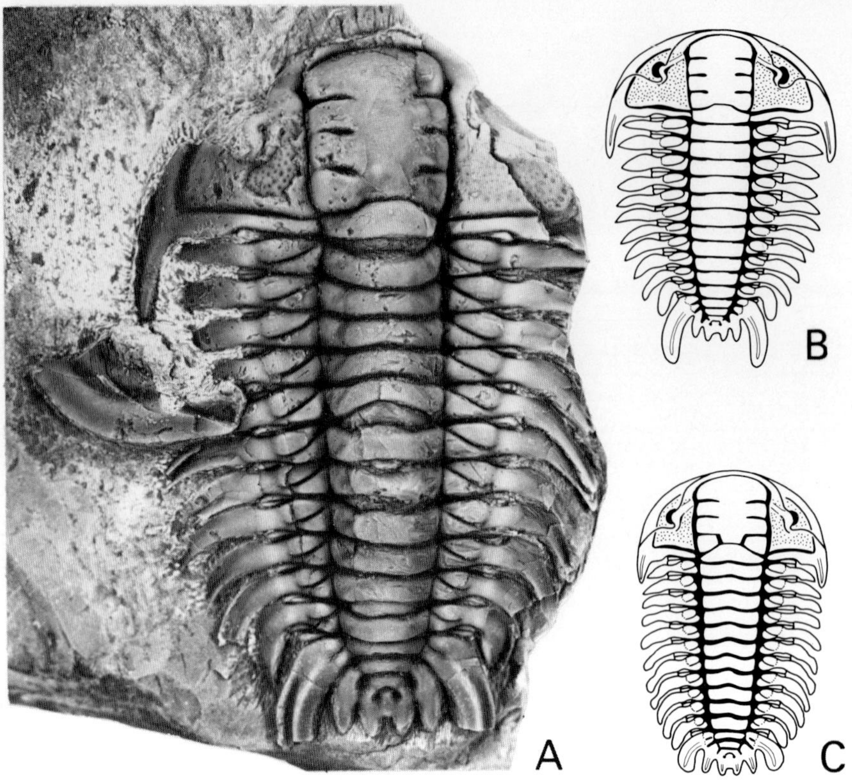

Fig. 35 ***Ceraurinus***
Ceraurinus marginatus Barton
A Exoskeleton, × 1.4, ROM 762 T, Cobourg Fm., Cobourg.
B Reconstruction, × .65.
Ceraurinus icarus (Billings)
C Reconstruction, × .9 based on UC 10837, Upper Ordovician, Indiana.

genus easily recognizable. Two pairs of distinct curving lateral glabellar furrows are present and the eyes are small and located far forward. The tiny pygidium is lenticular in outline. One species of *Triarthrus* has long and slender genal spines as well as a long median spine on the ninth thoracic segment. The slab illustrated in Figure 38 shows about 50 specimens of *Triarthrus*, each of which is missing its cranidium. I interpret these headless trilobites as moult remains, and feel that the moulting trilobite carried away the cranidium to be dropped elsewhere, perhaps in a locality like that shown in Figure 39A.

SPECIES *Triarthrus eatoni* (Hall) and *Triarthrus spinosus* Billings—Whitby Formation, southern Ontario and Ottawa area. The genus also occurs in the Lake Timiskaming area. Several other species have been named (*Triarthrus glaber, T. canadensis, T. huguesensis, T. spinosus rougensis, T. spinosus narrawayi*), but the relationship of these species to the first two species is unclear.

REFERENCES Parks (1928), Whittington (1957), Cisne (1973, 1975)

Fig. 36 ***Pseudogygites***
Pseudogygites latimarginatus (Hall)
Exoskeleton, × 3, ROM 35029, lower Whitby Fm., Craigleith.

Fig. 37 ***Pseudogygites***
Pseudogygites latimarginatus (Hall)
Bedding plane with numerous specimens, × .9, ROM 34519, lower Whitby Fm., Craigleith.

Fig. 38 ***Triarthrus***
Triarthrus eatoni (Hall)
Slab with about 50 moult remains of the trilobite and four large straight cephalopods, × .9, ROM 35375, lower Whitby Fm., Craigleith (?).

Fig. 39 ***Triarthrus***

Triarthrus eatoni (Hall)

A Slab with numerous cranidia, × 1.8, ROM 35376, lower Whitby Fm., Craigleith.

B Exoskeleton, × 2.8, ROM 35377, lower Whitby Fm., Craigleith.

Triarthrus spinosus Billings

C Slab with four dismembered exoskeletons, × 1.8, ROM 35378, Whitby Fm., Ottawa.

Silurian Trilobites

Figs. 40 to 48

Only about half as many trilobite genera occur in Silurian rocks of southern Ontario as occur in Ordovician rocks (Fig. 15). The Cataract Group, exposed at the base of the Niagara Escarpment from Niagara Falls to Manitoulin Island, contains at least eight genera of trilobites—most of which are rather poorly preserved. The best-preserved and most diverse Silurian trilobite faunas of southern Ontario occur in the Amabel Formation and its correlatives, the Rochester and Lockport Formations. In Silurian rocks, as in the Ordovician rocks, trilobites are never the dominant faunal element. From the Rochester Formation on the Niagara Peninsula, Bolton (1957) recorded 139 species of fossils, only six of which were trilobites. Trilobites are preserved as external and internal moulds in the Lockport Formation. A sparse, but well-preserved trilobite assemblage is found in the Eramosa Formation in the Hamilton area. The youngest Silurian formations in southern Ontario, the Salina, Bass Island, and Bertie Formations, have not yielded trilobites. The Bertie, however, has yielded spectacular specimens of that distant relative of the trilobite—the eurypterid.

A fossiliferous Silurian succession in the Lake Timiskaming area contains well-preserved trilobites. Bolton and Copeland (1972) illustrated species of *Otarion, Diacalymene, Encrinurus, Cheirurus, Acernaspis, Scutellum,* and *Ekwanoscutellum* from the Thornloe Formation of this area.

Further to the north in the Hudson Bay area, Silurian rocks are widely exposed. The few trilobites recorded from these rocks include *Encrinurus, Bumastus, Calymene,* and *Ekwanoscutellum* from the Severn River, Ekwan River, and Attawapiskat Formations.

***Acernaspis* Campbell Fig. 40A,B**

A few small, poorly preserved trilobites from the Cataract Group and Thornloe Formation are probably this early phacopid genus. *Acernaspis* has a forwardly expanding and weakly inflated glabella with three pairs of thin, curving lateral furrows. The eyes are relatively large and similar to those of *Phacops* (Fig. 57B). The small lenticular pygidium has a few furrows. The dorsal surface is very finely granulate.

SPECIES *Acernaspis* spp.—Cataract Group, Niagara Peninsula and Thornloe Formation, Lake Timiskaming.

REFERENCE Campbell (1967)

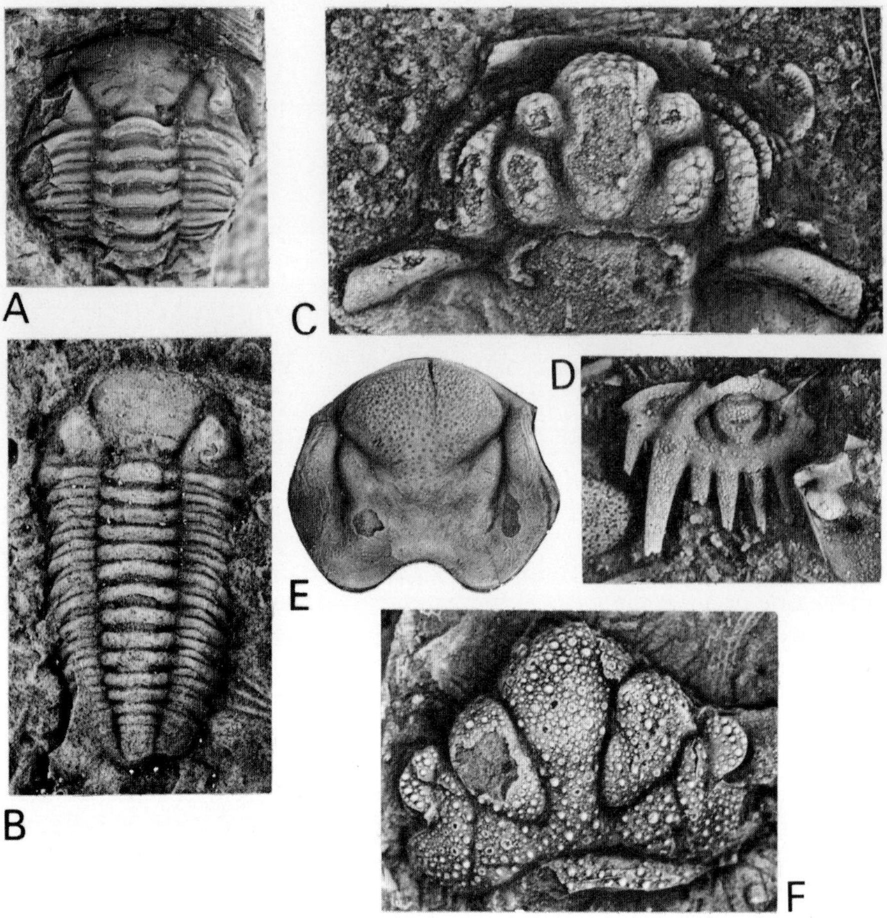

Fig. 40 ***Acernaspis Leonaspis Dicranopeltis***

Acernaspis spp.

A Enrolled specimen, × 3, GSC 20496, Grimsby Fm., DeCew Falls.

B Exoskeleton, × 2.6, ROM 303, Cataract Gp., Hamilton.

Leonaspis cf. *illinoiensis* (Weller)

C Cranidium, × 4.9, ROM 665, Manitoulin Fm., Credit Forks.

D Pygidium, × 4.9, ROM 442 N, Cataract Gp., Eugenia.

Dicranopeltis fragosa Phleger

E Hypostome, × 2.6, ROM 35379, Rochester Fm., Grimsby.

F Cranidium, × 3.4, ROM 35380, Rochester Fm., Grimsby.

Leonaspis Richter and Richter Fig. 40C,D

Leonaspis has a round glabella with two pairs of inflated lateral lobes. The glabella of this odontopleurid trilobite is flanked by a pair of long narrow palpebral lobes. The lateral margin of the cephalon is ornamented by a row of spines. The pygidium is semicircular in outline and bears three or four pairs of pointed unequally sized spines.

SPECIES *Leonaspsis* cf. *illinoiensis* (Weller)—Cataract Group, southern Ontario.

REFERENCES Whittington (1956), Bruton (1968), Campbell (1977)

***Dicranopeltis* Hawle and Corda Fig. 40E,F**

This rare lichid trilobite is distinguished chiefly by its glabellar lobation which comprises a median anchor-shaped lobe flanked by a pair of elliptical lateral lobes. A pair of small triangular lobes are located immediately in front of the occipital furrow. The palpebral lobes are large and located close to the axial furrow.

SPECIES *Dicranopeltis fragosa* Phleger—Rochester Formation, Niagara Peninsula.

REFERENCES Weller (1909), Phleger (1937), Warburg (1939)

***Encrinurus* Emmrich Figs. 41, 42**

The glabella of this encrinurid trilobite expands forward and has three pairs of obscure lateral furrows. The cephalon is semicircular in outline, arched, and densely covered by large perforated tubercles. The helmet-shaped hypostome

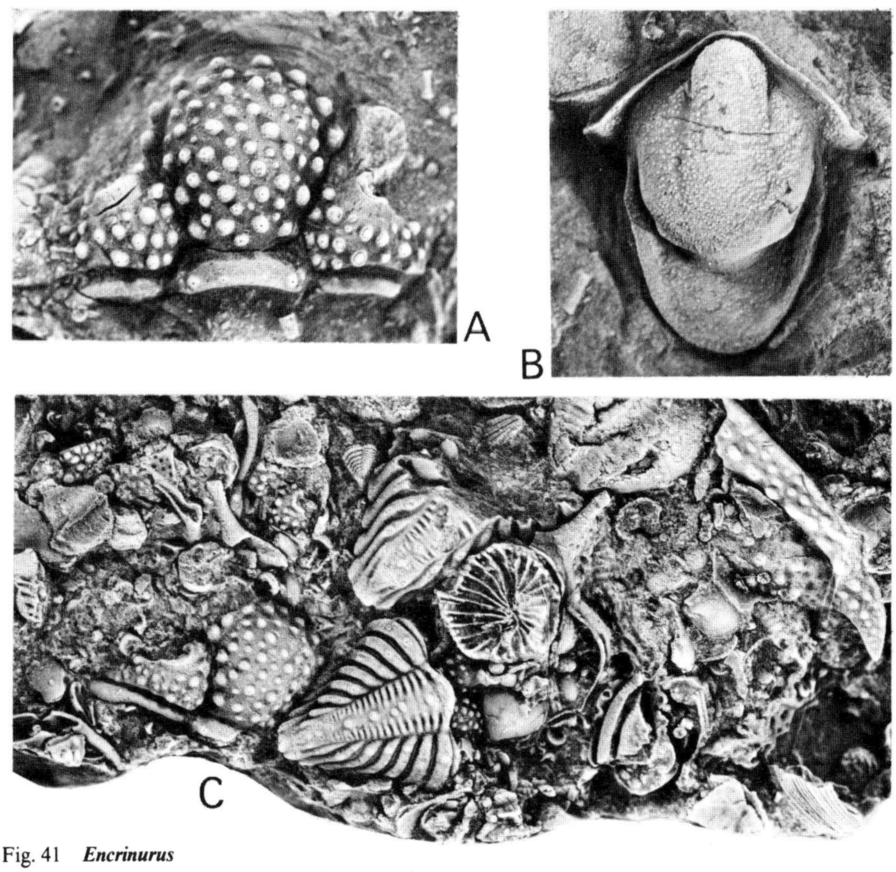

Fig. 41 ***Encrinurus***
Encrinurus ornatus Hall and Whitfield
A Cranidium, × 2.6, McM-S345, Eramosa Fm., Dundas.
B Hypostome, × 4.7, McM-S343, Eramosa Fm., Dundas.
C Shell hash including a well-preserved pygidium, × 2.3, McM-S348, Eramosa Fm., Dundas.

Fig. 42 ***Encrinurus***
Bedding plane crowded with cranidia, pygidia, hypostomes, and free cheeks of *Encrinurus ornatus*. Also occurring are fragments of *Calymene* and *Dalmanites*, as well as brachiopods and corals, × .9, McM-S353, Eramosa Fm., Dundas.

consists of an inflated anterior portion and a flat tonguelike posterior brim. The acutely triangular pygidium consists of a densely ribbed axis with a median row of tubercles and a pleural portion of about eight curving ribs.
SPECIES *Encrinurus ornatus* Hall and Whitfield, *Encrinurus deltoidens* Shumard, and *Encrinurus tuberculifrons* Weller—Lower and Middle Silurian rocks, southern Ontario and Hudson Bay area.
REFERENCES Best (1961), Tripp (1962), Tripp *et al.* (1977)

***Dalmanites* Barrande Fig. 43A**
Dalmanites has a glabella of low convexity that expands slightly towards the front. The glabella has three pairs of lateral furrows and it is flanked by relatively large eyes. The cephalon has a distinct flattened border and long tapering genal spines. The pygidium is triangular and finely ribbed and may terminate in a long slender spine. This dalmanitid trilobite is relatively common in a number of Lower and Middle Silurian formations in southern Ontario, but complete specimens are essentially confined to the shales of the Rochester Formation.
SPECIES *Dalmanites limulurus* (Green)—Lower and Middle Silurian rocks, southern Ontario.
REFERENCES Delo (1940), Campbell (1967), Whittington and Campbell (1967)

***Trimerus* Green Fig. 43B,C**
Trimerus is a large inflated homalonotid trilobite with a triangular cephalon and small eyes. Its glabella narrows slightly towards the front. The axis on the thorax is very wide and only faintly outlined by furrows. The axial furrows are distinct on the pygidium, which is segmented and tapered to a sharp point.
SPECIES *Trimerus delphinocephalus* Green—Rochester Formation, Niagara Peninsula.
REFERENCES Tomczykowa (1975), Thomas (1977)

***Arctinurus* Castelnau Fig. 44**
In 1825 Dr. J.J. Bigsby, who five years earlier had collected the first North American trilobite to receive a name, described a new trilobite *Paradoxus boltoni* from shales exposed at Lockport, New York. This species became the type of *Arctinurus*—certainly one of the more striking and impressive of the Silurian trilobites. *Arctinurus* is a large, ovate, and rather flat lichid trilobite with a very broad axis and a distinctive tongue-shaped prow in front of the cephalon. The glabella is much narrower than the thoracic axis and is divided into three lobes by a pair of hook-shaped longitudinal furrows. The doublure is very broad and flat. The pygidium is large and quadrate and carries three pairs of lobate spines. Articulated specimens of *Arctinurus* are very rare, but fragments of this trilobite are not uncommon in the Rochester Formation.
SPECIES *Arctinurus boltoni* (Bigsby)—Rochester Formation, Niagara Peninsula.
REFERENCES Phleger (1937), Weller (1909)

***Calymene* Brongniart Figs. 45C,D, 50C,D**
The name *Calymene* was formerly a catch-all for most calymenid trilobites, but largely as a result of the work of Shirley (1936) and Whittington (1971) we now have a more realistic classification of these trilobites. Under this classification,

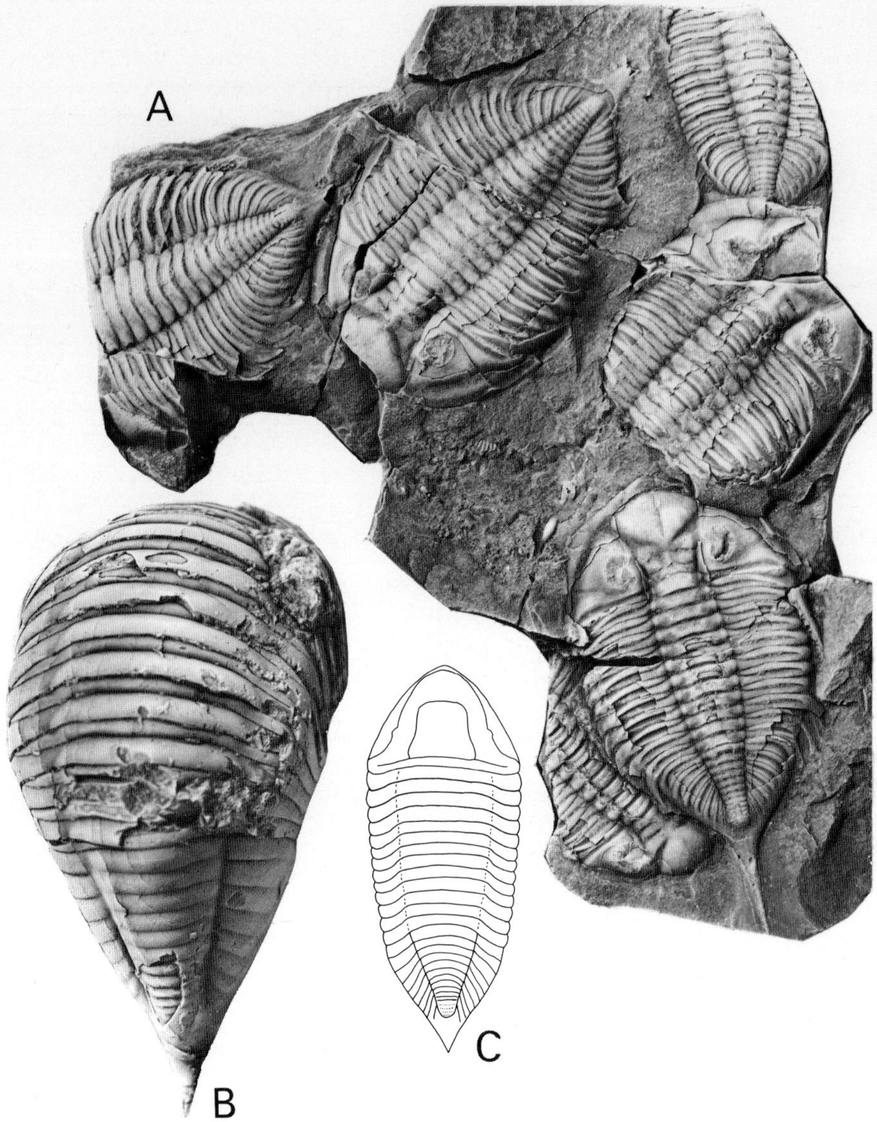

Fig. 43 ***Dalmanites Trimerus***
Dalmanites limulurus (Green)
A Bedding plane with six individuals, × 1.2, ROM 35683, Rochester Fm., St. Catharines.
Trimerus delphinocephalus Green
B Thorax and pygidium, × 1.1, ROM 29634, Rochester Fm., DeCew Falls.
C Reconstruction, × .4.

the name *Calymene* is restricted to calymenid trilobites with an inflated glabella, a narrow and deep preglabellar furrow, and a papillate second lateral glabellar lobe (that is, a narrow buttress extending part way across the axial furrow opposite the eye). The pygidium is nearly identical to that of *Flexicalymene*.

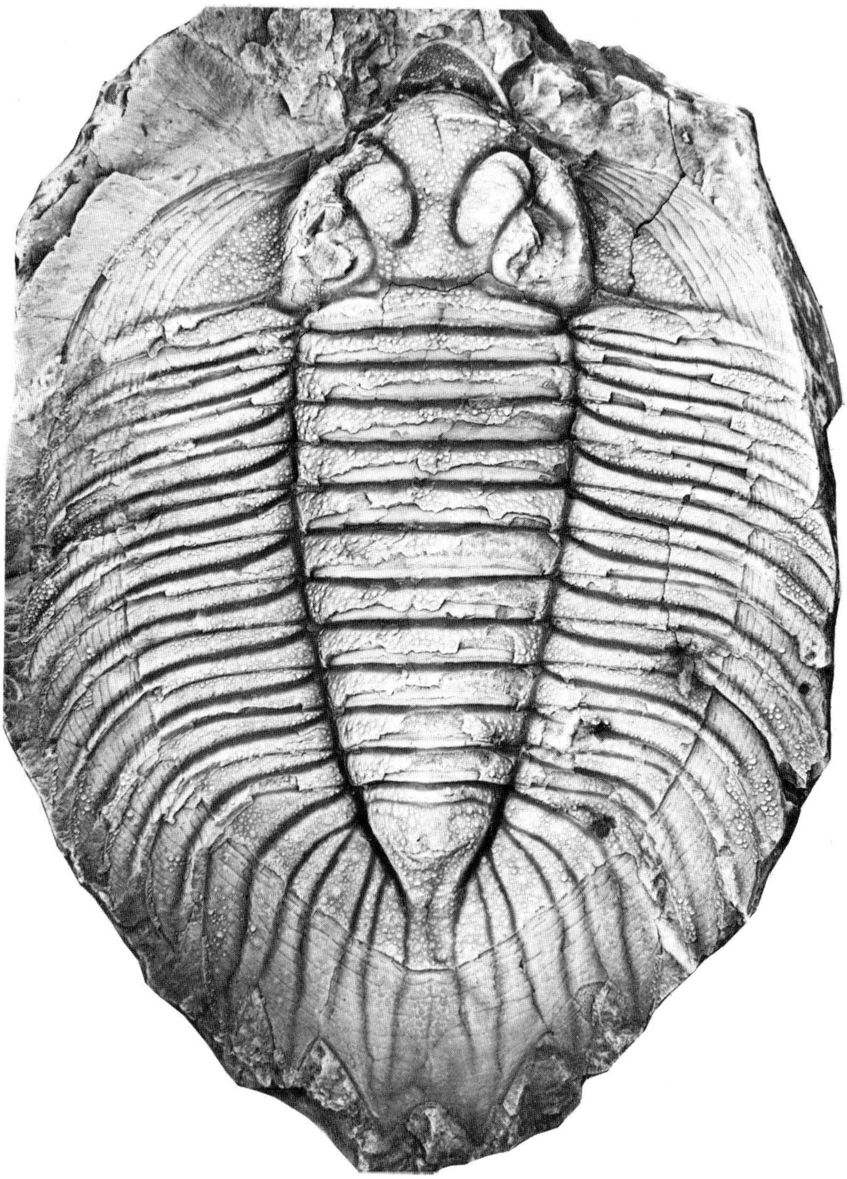

Fig. 44 ***Arctinurus***
Arctinurus boltoni (Bigsby)
Exoskeleton, × 1, NYSM 9631, Rochester Fm., Lockport, New York.

SPECIES *Calymene niagarensis* (Hall) and *Calymene breviceps* Raymond—Cataract, Amabel, and Rochester Formations, southern Ontario. *Calymene platys* Green—Bois Blanc Formation, southern Ontario.
REFERENCES Shirley (1936), Campbell (1967), Whittington (1971)

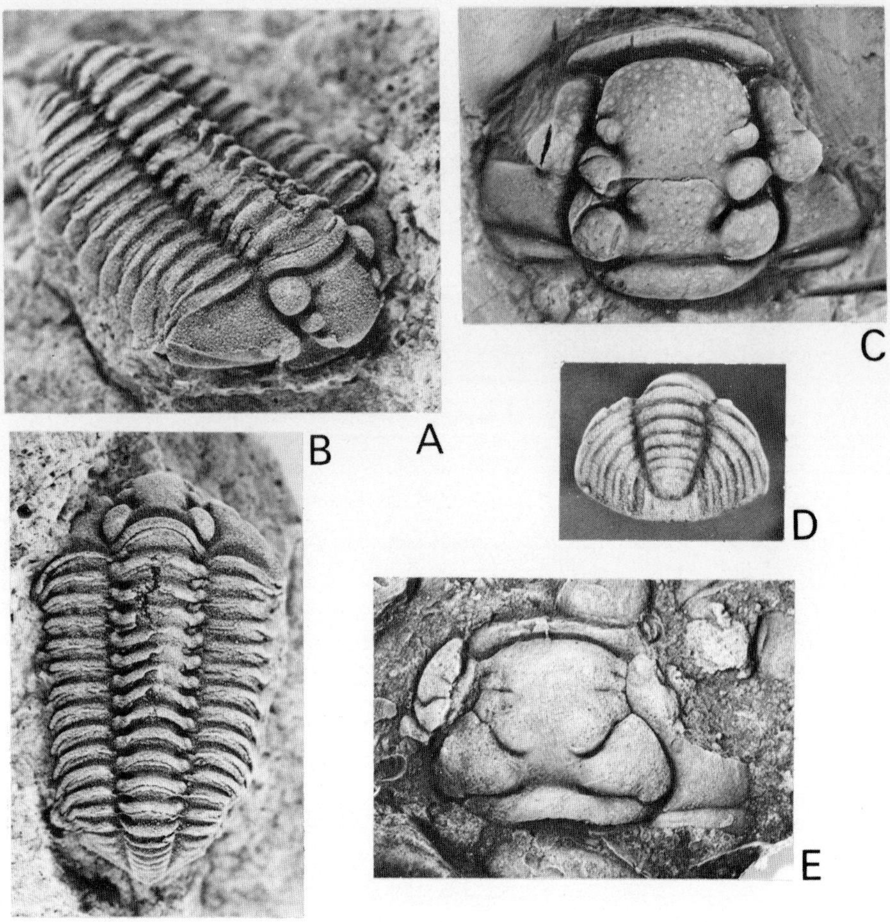

Fig. 45 ***Flexicalymene Calymene Liocalymene***
Flexicalymene celebra (Raymond)
A,B Exoskeleton, × 2, ROM 580 N, Lockport Fm., Hamilton.
Calymene breviceps Raymond
C Cranidium, × 2.8, McM-S572a, Eramosa Fm., Dundas.
D Pygidium, × 2.8, McM-S572b, Eramosa Fm., Dundas.
Liocalymene clintoni (Vanuxem)
E Cranidium, × 1.9, USNM 154493, Crab Tree Orchard Gp., Tolsboro, Kentucky.

***Liocalymene* Raymond Fig 45E**

This calymenid trilobite has a bell-shaped glabella with very large first glabellar lobes. The cranidium is rather flat and the glabellar lobes are much less inflated than those of *Flexicalymene* or *Calymene*. The pygidial axis is furrowed, the remainder of the pygidium is smooth.

SPECIES *Liocalymene clintoni* (Vanuxem)—Cataract Group and Rochester Formation, southern Ontario.

REFERENCE Whittington (1971)

***Scutellum* Pusch Fig. 46A,B**

Scutelluid trilobites from Silurian rocks of North America are imperfectly known from isolated cranidia and pygidia. The two specimens illustrated here are assigned to the genus *Scutellum* although it is possible that they belong to other scutelluid trilobites such as *Kosovopeltis* or *Planiscutellum*. The cranidium is lowly vaulted and possesses a narrow anterior border. The glabella expands forward from its narrowest portion in front of the occipital furrow. The axial furrow and three pairs of lateral glabellar furrows are shallow. The pygidium is nearly flat and semi-elliptical in outline. It has seven pairs of shallow furrows which radiate from a short triangular axis.

SPECIES *Scutellum acamus* (Hall)—Amabel Formation, *Scutellum rochesterense* Howell and Sanford—Irondequoit and Amabel Formations, southern Ontario.

REFERENCE Šnajdr (1960)

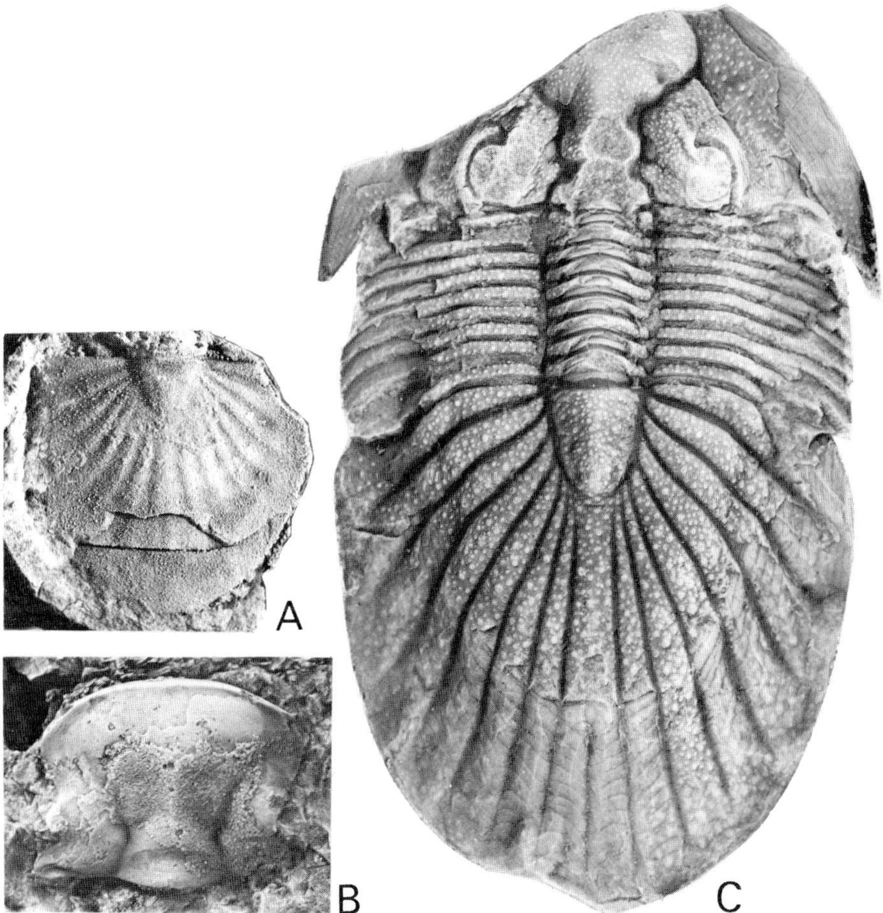

Fig. 46 ***Scutellum Ekwanoscutellum***

Scutellum acamus (Hall)

A Pygidium, × 1.1, GSC 20595, Amabel Fm., Owen Sound.

Scutellum rochesterense Howell and Sanford

B Cranidium, × 1.9, PU 57696, Irondequoit Fm., Rochester, New York.

Ekwanoscutellum ekwanensis (Whiteaves)

C Exoskeleton, × 1.2, ROM 28075, Ekwan River Fm., Owl River, Manitoba (Hudson Bay area).

Ekwanoscutellum Přibyl and Vaněk Fig 46C

This scutelluid trilobite is characterized by a moderately inflated and large semi-elliptical pygidium with seven paired furrows that curve back from the short axis almost to the margin of the pygidium. A shorter unpaired median furrow does not reach the axis. The goblet-shaped and inflated glabella is well-defined by deep furrows. It has two pairs of deep lateral furrows that seem to constrict the posterior part of the glabella and a third pair represented by a pit. The borders of the cephalon are broad and flat and the genal spines are short and pointed. The entire dorsal surface is covered by coarse granules.

SPECIES *Ekwanoscutellum ekwanensis* (Whiteaves)—Ekwan River Formation, Hudson Bay area and *Ekwanoscutellum lamphami* (Whitfield)—Amabel Formation, Manitoulin Island and Lake Timiskaming.

REFERENCES Fritz (1964), Přibyl and Vaněk (1971)

Cheirurus Beyrich Fig. 47A–C

Cheirurus has an inflated glabella that expands towards the front where it overhangs the anterior border. The lateral glabellar furrows are deep and straight and the basal glabellar lobe is isolated. The eye is located opposite the second glabellar furrow. The pygidium bears three pairs of moderately long, curved, and pointed spines and a single median spine.

SPECIES *Cheirurus niagarensis* (Hall) and *Cheirurus welleri* Raymond—Amabel Formation, southern Ontario and Lake Timiskaming.

REFERENCES Raymond (1916), Weller (1909), Campbell (1967), Lane (1971)

Bumastus Murchison Figs. 47D, 48B,C

Bumastus is a large illaenid trilobite with an oval outline and a very broad axis. The smooth semicircular cephalon is characterized by widely spaced eyes and very faint medially-bowed axial furrows that fade towards the front. The axial furrows are not discernable on the thorax nor on the smooth semicircular pygidium.

SPECIES *Bumastus ioxus* (Hall)—Cabot Head, Rochester, Amabel, and Guelph Formations and *Bumastus aboynensis* (Whiteaves)—Guelph Formation, southern Ontario.

REFERENCES Weller (1909), Raymond (1916)

Sphaerexochus Beyrich Fig. 48A

This cheirurid trilobite has a round and greatly inflated glabella. The cheeks are much reduced and nearly vertically disposed beneath the glabella. The only evident glabellar furrow is a strongly curved first glabellar furrow that isolates a circular first glabellar lobe. The pygidium bears three pairs of short, blunt spines.

SPECIES *Sphaerexochus romingeri* Hall—Amabel Formation, southern Ontario.

REFERENCES Lane (1971), Chatterton and Ludvigsen (1976), Perry and Chatterton (1977)

Fig. 47 ***Cheirurus Bumastus***
Cheirurus niagarensis (Hall)
A Three cranidia, × 1.5, ROM 551 N, Niagaran Dolomite, Chicago, Illinois.
B Cranidium, × 1, GSC 17969, Lockport Fm., Grimsby.
C Pygidium, × 1, GSC 17990, Lockport Fm., Grimsby.
Bumastus aboynensis (Whiteaves)
D Cephalon, × 1.5, GSC 3016, Guelph Fm., Aboyne near Elora.

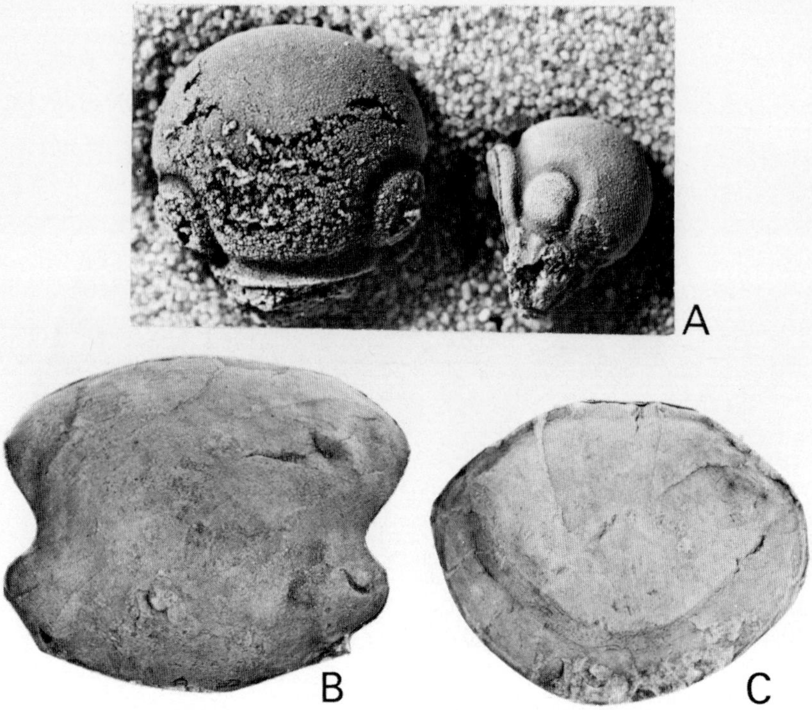

Fig. 48 ***Sphaerexochus Bumastus***
Sphaerexochus romingeri Hall
A Two cranidia, × 1.7, ROM 1345 N and ROM 552 N, Racine Dolomite, Wisconsin.
Bumastus ioxus (Hall)
B Cranidium, × .7, ROM 1540a N, Lockport Fm., Hamilton.
C Pygidium, × .7, ROM 1540b N, Lockport Fm., Hamilton.

Devonian Trilobites

Figs. 49 to 58

In southern Ontario occurrences of Devonian trilobites are essentially confined to the Bois Blanc, Amherstburg, and Dundee Formations and to the Hamilton Group. Well-preserved specimens of trilobites, including the rare and spectacular *Terataspis*, occur in the Bois Blanc Formation along the southern portion of the Niagara Peninsula, but are perversely difficult to collect because of the toughness of the rock. The Formosa Reef of the Amherstburg Formation is exposed in and around the village of Formosa and contains sparse, but well-preserved trilobite assemblages. The trilobites of the Dundee Formation occur in a tough dense limestone and are difficult to extract. The shales and limestones of the Hamilton Group of southwestern Ontario yield well-preserved and moderately abundant trilobites, including *Phacops* and *Greenops*, but trilobites are a distinct minority among the diverse faunas of this unit. Stumm and Wright (1958) listed 341 species of fossils from the Hungry Hollow Formation (middle Hamilton Group) near Thedford, Ontario, only five of which were trilobites.

The Devonian rocks of the Hudson Bay area contain moderately rich, but unstudied, trilobite faunas. Sanford and Norris (1975) recorded *Dalmanites, Phacops, Proetus*, and *Synphoria* from the Stooping River Formation and *Dalmanites, Dechenella, Odontocephalus, Odontochile, Phacops, Proetus*, and *Reedops* from the Kwataboahegan Formation.

***Terataspis* Hall Figs. 49, 50A,B**

Because of its size and ornamentation, *Terataspis* is probably the most impressive trilobite found in Ontario. Complete specimens are not known, but a fragment of a cranidium that must have belonged to an individual in excess of 50 cm in length was found in southern Ontario (Hall and Clarke, 1888). This lichid trilobite is therefore one of the largest trilobites ever found. *Terataspis* is characterized by the swollen front portion of the median glabellar lobe which is undercut by backwardly converging longitudinal furrows. These furrows delineate a pair of swellings located in line with the eyes. The pygidium bears four pairs of slightly curved and barbed spines. The entire dorsal surface is covered by spines and tubercles of varying sizes. The kidney-shaped hypostome is nearly flat and has a poorly defined central body.

SPECIES *Terataspis grandis* (Hall)—Bois Blanc Formation, Niagara Peninsula.

REFERENCES Hall and Clarke (1888), Reiman (1945)

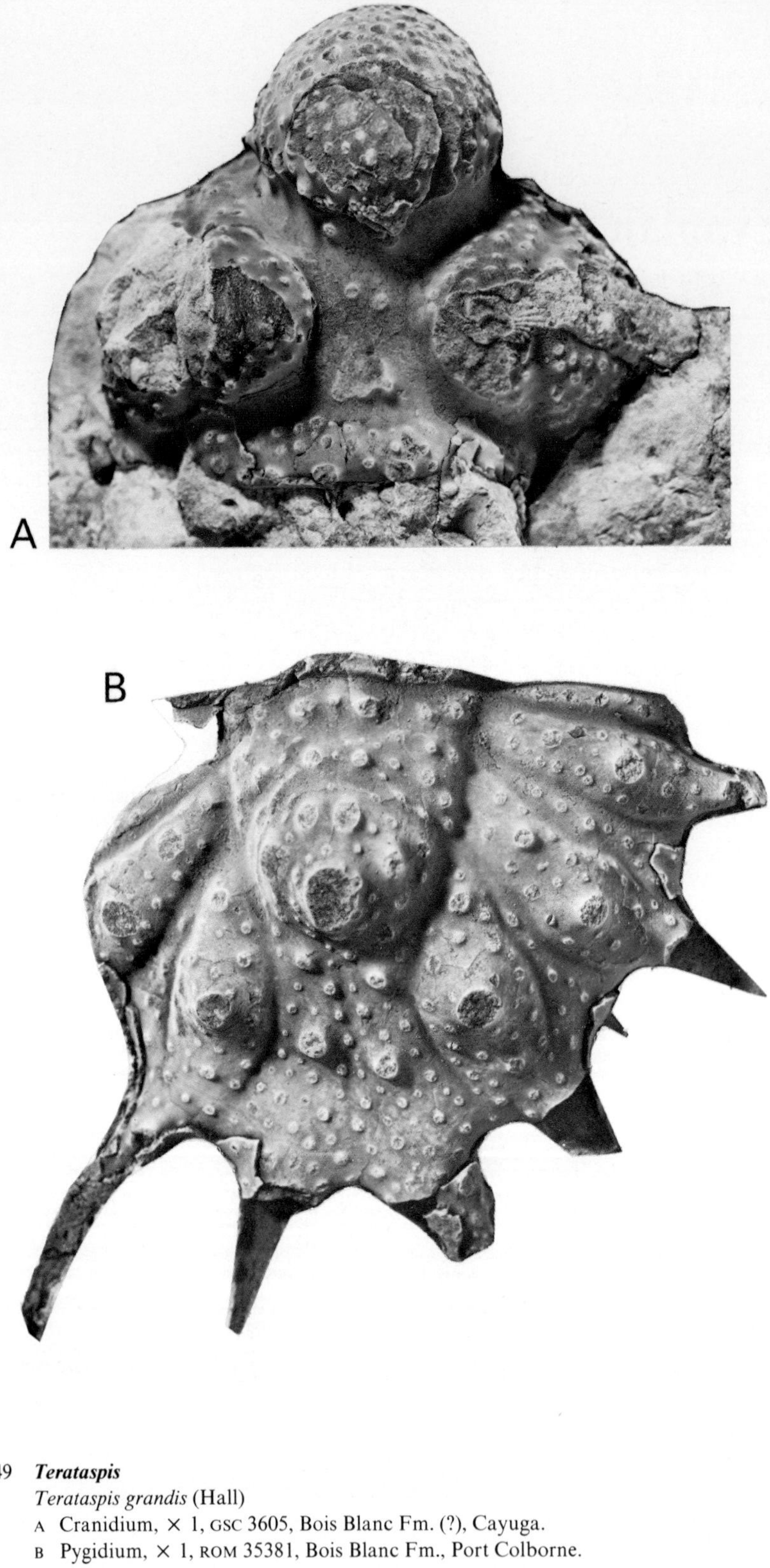

Fig. 49 ***Terataspis***

Terataspis grandis (Hall)

A Cranidium, × 1, GSC 3605, Bois Blanc Fm. (?), Cayuga.

B Pygidium, × 1, ROM 35381, Bois Blanc Fm., Port Colborne.

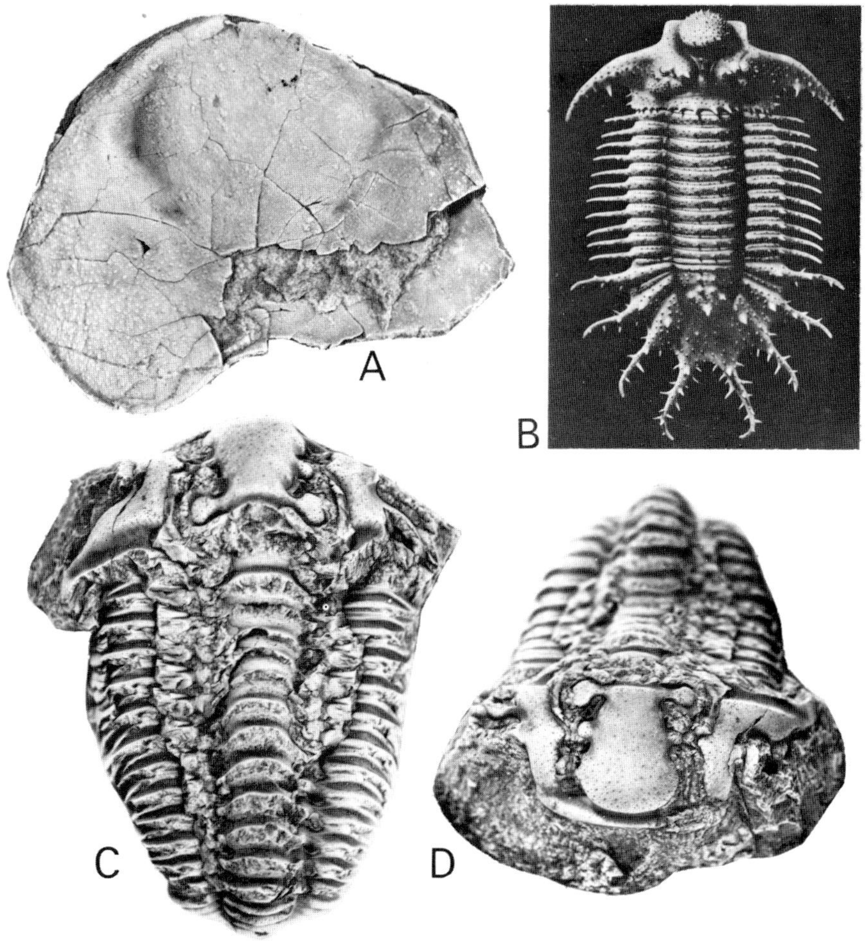

Fig. 50 ***Terataspis Calymene***
Terataspis grandis (Hall)
A Hypostome, × .9, ROM 35396, Bois Blanc Fm., Port Colborne.
B Reconstruction, × .16, copy from Reimann (1945).
Calymene platys Green
C,D Exoskeleton, × .9, ROM 874 UH, Bois Blanc Fm., Hagersville.

Anchiopsis Delo Fig. 51A,B

Anchiopsis is relatively common in the Bois Blanc Formation, but is invariably represented by fragmentary material. This dalmanitid trilobite has two pairs of short, lateral glabellar furrows behind a longer furrow. The innermost tips of these furrows are joined by faint longitudinal furrows. A short, stout spine occurs on the occipital ring. The triangular pygidium has seven to nine ribs and an upwardly and backwardly directed median posterior spine.

SPECIES *Anchiopsis anchiops* (Green)—Bois Blanc Formation, southern Ontario.

REFERENCES Hall and Clarke (1888), Delo (1940), Stumm (1954), Lespérance and Bourque (1971)

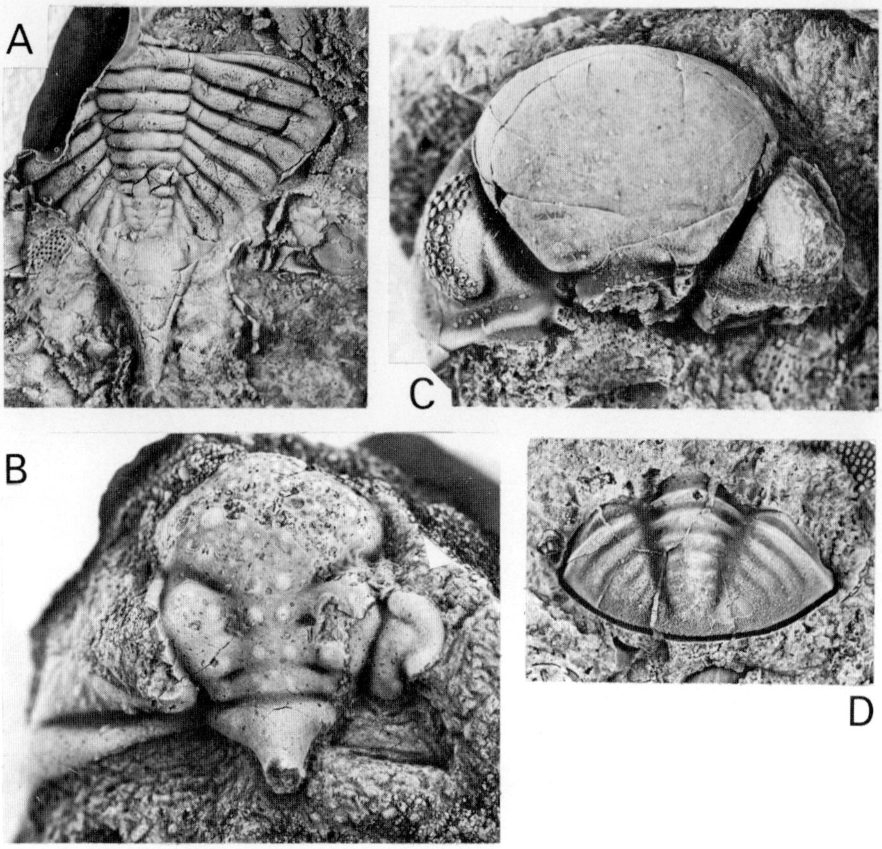

Fig. 51 ***Anchiopsis Phacops***
Anchiopsis anchiops (Green)
A Pygidium, × 2, ROM 35382, Bois Blanc Fm., Hagersville.
B Cranidium, × 2, ROM 35383, Bois Blanc Fm., Hagersville.
Phacops cristata Hall
C Cephalon, × 2, ROM 35384, Bois Blanc Fm., Hagersville.
D Pygidium, × 2, ROM 35385, Bois Blanc Fm., Hagersville.

***Phacops* Emmrich Figs. 51C,D, 56, 57A,B**

Phacops is probably the most common Devonian trilobite in Ontario. It is also one of the best-known of all trilobites because it is robust and generally well preserved; it has attracted the attention of many palaeontologists over the years. The semicircular cephalon of *Phacops* is dominated by a large, forwardly expanding and inflated glabella that is covered by closely spaced and large tubercles; it is outlined by deep axial furrows. Only the first pair of lateral glabellar furrows are clearly evident; the remaining two pairs are largely obscured by the tubercles. The aggregate eye is large and consists of a number of domical lenses, each of which is clearly separated from its neighbour. The lenses are arranged in as many as 18 vertical rows, each containing up to nine lenses. The pygidium is lenticular in outline and has a well-defined axis and four to six furrows.

Besides being a very well-preserved individual of *Phacops rana*, the enrolled specimen illustrated in Figure 56 is unique in having a withered left eye that has only 20 lenses compared with 67 on the normal right eye. The fewer number of lenses probably indicates an injury inflicted during an early period in the animal's life. In spite of its handicap, the trilobite must have managed quite well —judging from its size. But one can't help but wonder if this cock-eyed trilobite spent its life wandering about in sinistral circles.

SPECIES *Phacops cristata* Hall—Bois Blanc Formation, *Phacops rana* (Green)—Dundee Formation and Hamilton Group, *Phacops iowensis* Delo—Hamilton Group, southern Ontario. Each species includes a number of subspecies (Eldredge 1972, 1973).

REFERENCES Delo (1940), Stumm (1953a, 1954), Clarkson (1967), Campbell (1967, 1975b, 1977), Eldredge (1972, 1973), Kesling and Chilman (1975), Miller (1976), Chlupáč (1977)

***Acanthopyge* Hawle and Corda Fig. 52A,B**

Acanthopyge is a probable descendant of *Hemiarges* (Fig. 33) and shares a number of cranidial and pygidial features with that lichid genus. *Acanthopyge* has three glabellar lobes of equal convexity arranged in a transversely oval field on the front part of the cranidium. The narrow area between the central glabellar lobe and the occipital ring is depressed. The axial furrows are obsolete behind the lateral glabellar lobes. The lateral portions of the cranidium are inflated. The pygidium is semicircular, little inflated except for the axis, and carries three pairs of widely separated and pointed spines.

SPECIES *Acanthopyge contusa* (Hall and Clarke)—Amherstburg Formation (Formosa Reef), southern Ontario.

REFERENCES Hall and Clarke (1888), Tripp (1957), Fagerstrom (1961)

***"Dechenella"* Fig. 52C**

This small peculiar pygidium is semicircular in outline with rather steep margins. It has deep axial furrows outlining a segmented axis that extends nearly to the margin. The pygidium carries nine ribs—each of which terminates against a small rounded node located on the border. The rest of the exoskeleton is not known.

SPECIES *"Dechenella" halli* Stumm—Amherstburg Formation (Formosa Reef), southern Ontario.

REFERENCES Hall and Clarke (1888), Stumm (1953b), Fagerstrom (1961)

***Mystrocephala* Whittington Fig. 52D,E**

The cephalon of this small brachymetopid trilobite is crescentic and has concave borders whose interior portions are crowded with small tubercles. The convex glabella is parabolic in outline and has two pairs of lateral furrows. In front of the eyes, the facial sutures diverge markedly. The semicircular pygidium has a convex axis and seven segments—each of which bears an accessory rib that extends to the margin.

SPECIES *Mystrocephala stummi* Fagerstrom—Amherstburg Formation (Formosa Reef), southern Ontario.

REFERENCES Whittington (1960), Fagerstrom (1961)

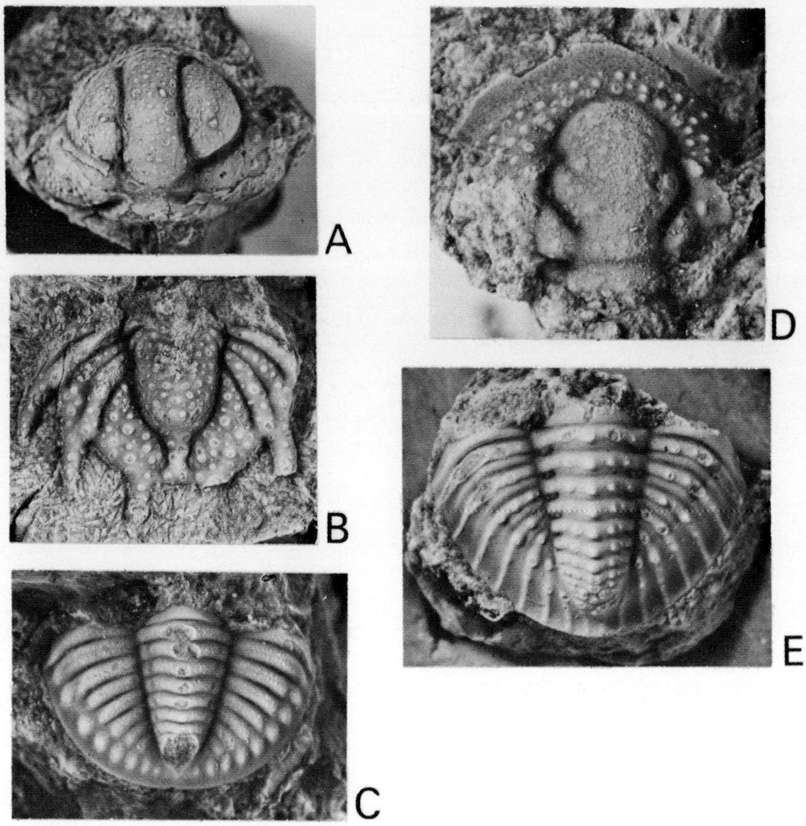

Fig. 52 ***Acanthopyge "Dechenella" Mystrocephala***
Acanthopyge contusa (Hall and Clarke)
A Cranidium, × 1.7, UMMP 36172, Amherstburg Fm. (Formosa Reef), Formosa.
B Pygidium, × 1.7, UMMP 36148, Amherstburg Fm. (Formosa Reef), Formosa.
"Dechenella" halli Stumm
C Pygidium, × 6.9, ROM 35397, Amherstburg Fm. (Formosa Reef), Formosa.
Mystrocephala stummi Fagerstrom
D Cranidium, × 6.9, ROM 35398, Amherstburg Fm. (Formosa Reef), Formosa.
E Pygidium, × 6.9, ROM 35399, Amherstburg Fm. (Formosa Reef), Formosa.

Coronura Hall and Clarke Fig. 53

The only relatively complete specimen of this rare trilobite (Fig. 53A) was collected in New York, but the genus has been widely reported from the lower Middle Devonian rocks of Ontario. *Coronura* is one of the largest dalmanitid trilobites known. According to Hall and Clarke (1888), it reaches a length of nearly 50 cm. *Coronura* is characterized by a rather flat cephalon with broad borders, a forwardly expanding glabella, high conical eyes, a large pygidium with up to 18 ribs, a pair of large posteriorly directed spines, and a fringe of smaller spines along the lateral margins.

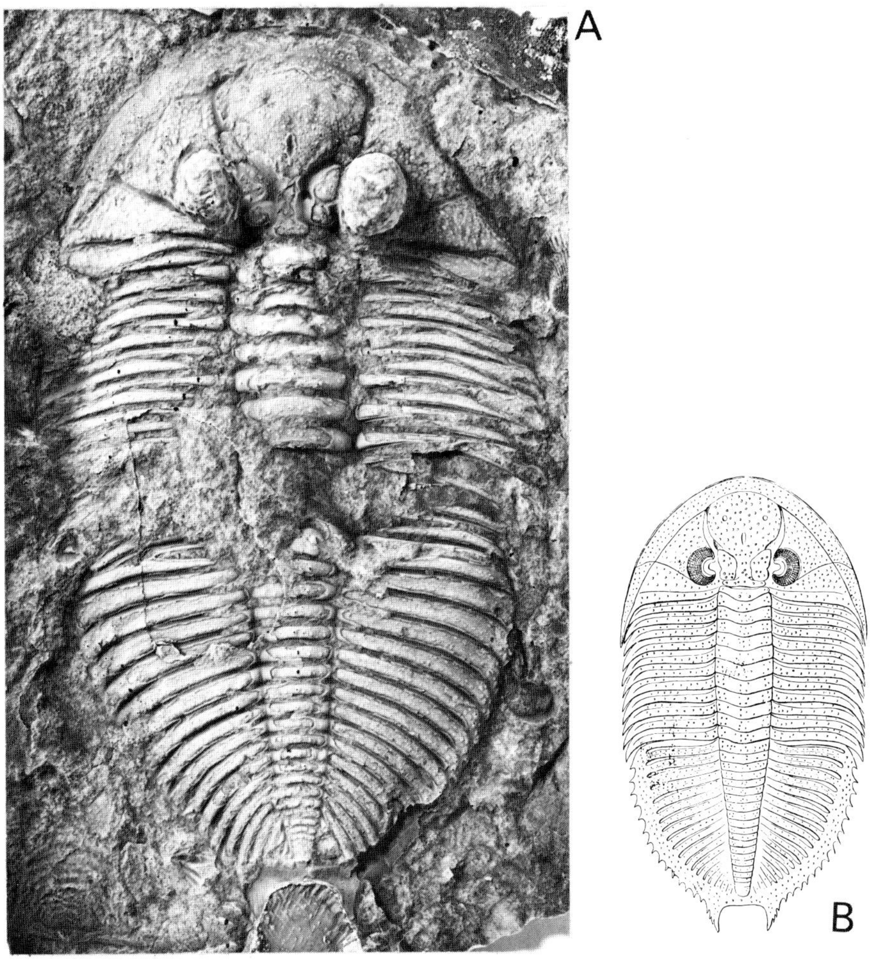

Fig. 53 ***Coronura***
Coronura aspectans (Conrad)
A Exoskeleton, × .9, NYSM 4316, Onondaga Fm., Leroy, New York.
B Reconstruction, × .4, copy from Clarke (1892).

SPECIES *Coronura aspectans* (Conrad) and *Coronura myrmecophorus* (Green)—Dundee Formation, southern Ontario.
REFERENCES Delo (1940), Stumm (1954), Lespérance and Bourque (1971), Lespérance (1975)

Trypaulites Delo Fig. 54A
Trypaulites is similar to *Coronura*, differing principally in possessing relatively large eyes and a pygidium that has an axis with a triangular cross-section and which lacks marginal spines.
SPECIES Although *Trypaulites* has not yet been reported from Ontario rocks, it is included here because it occurs in adjoining areas (New York, Ohio, Michi-

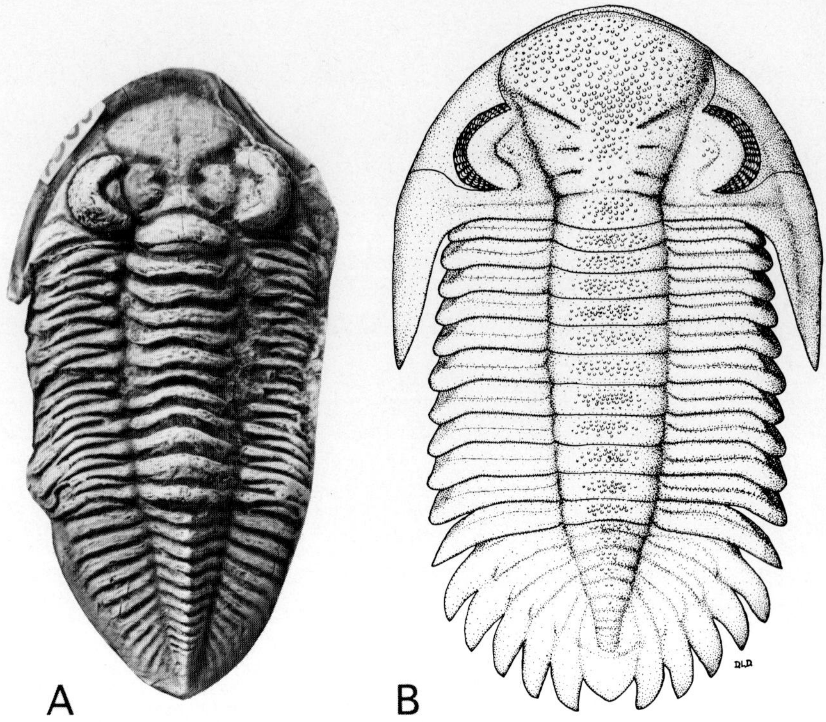

Fig. 54 ***Trypaulites* *Greenops***
Trypaulites calypso (Hall)
A Exoskeleton, × 1.1, NYSM 4305, Columbus Limestone, Sandusky, Ohio.
Greenops boothi (Green)
B Reconstruction, × 3.

gan, Quebec) in rocks correlative with the Bois Blanc, Amherstburg, and Dundee Formations.
REFERENCES Delo (1940), Stumm (1954), Lespérance (1975)

***Greenops* Delo Figs. 54B, 55A,B**
Greenops is a common dalmanitid trilobite in the shales and limestones of the Hamilton Group in southern Ontario and is easily recognized by its distinctive spiny pygidium. The crescent-shaped cephalon is characterized by a forwardly expanding glabella which has three pairs of narrow and straight lateral furrows —each of which terminates before reaching the axial furrow. The elevated eyes are large and strongly curved in a horizontal plane. The genal spines are long and blunt. The semicircular pygidium is bordered by a fringe of five pairs of slightly curved spines and a single unpaired spine.

Like most of the later inhabitants of eastern North America, *Greenops* was an immigrant—the ancestry of the genus appears to lie with Early Devonian trilobites of Europe and north Africa. A closely related genus, *Neometacanthus* (Fig. 57D), occurs in New York State.

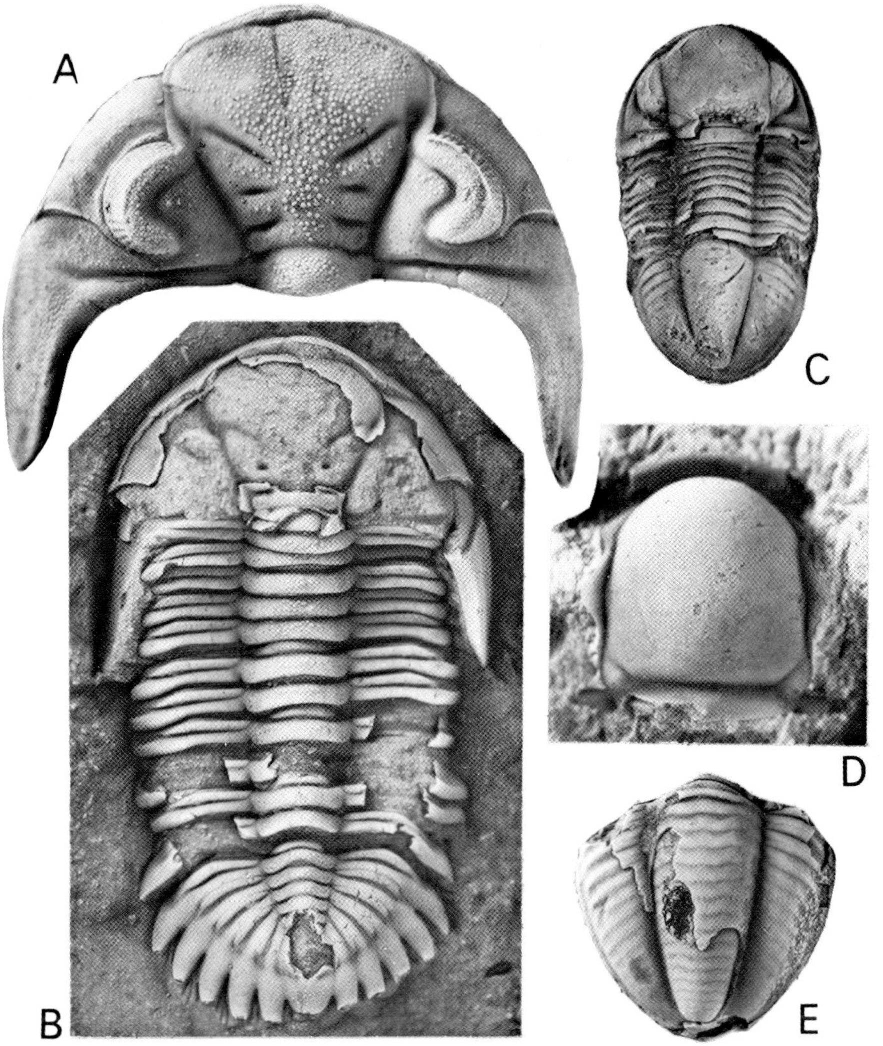

Fig. 55 ***Greenops Crassiproetus***

Greenops boothi (Green)

A Cephalon, × 4.3, ROM 1731 H, Hamilton Gp., Thetford.

B Exoskeleton, × 3, ROM 1066, Hamilton Gp., Thetford.

Crassiproetus canadensis Stumm

C Exoskeleton, × 1.9, UMMP 27090, Hamilton Gp., Arkona.

Crassiproetus crassimarginatus (Hall)

D Cranidium, × 2.6, ROM 35387, Amherstburg Fm. (Formosa Reef), Formosa.

E Pygidium, × 1.1, ROM 35386, Amherstburg Fm. (Formosa Reef), Formosa.

SPECIES *Greenops boothi* (Green), *Greenops arkonensis* Stumm—Hamilton Group, southern Ontario.
REFERENCES Delo (1940), Stumm (1953a)

***Crassiproetus* Stumm Fig. 55C–E**

Crassiproetus is one of the more common Devonian trilobites in Ontario. This proetid trilobite is found in nearly every Devonian formation and is particularly abundant in the Bois Blanc and Amherstburg (Formosa Reef) Formations. The inflated, egglike glabella of *Crassiproetus* has a single distinct lateral furrow which isolates a small triangular basal lobe. The anterior border is relatively broad and flat. The palpebral lobes are small and located immediately adjacent to the axial furrows. The genal angles terminate bluntly. The highly inflated pygidium has a long and convex axis with 12 to 15 faint rings. The borders of the pygidium descend steeply.
SPECIES *Crassiproetus crassimarginatus* (Hall)—Bois Blanc, Amherstburg, Lu-

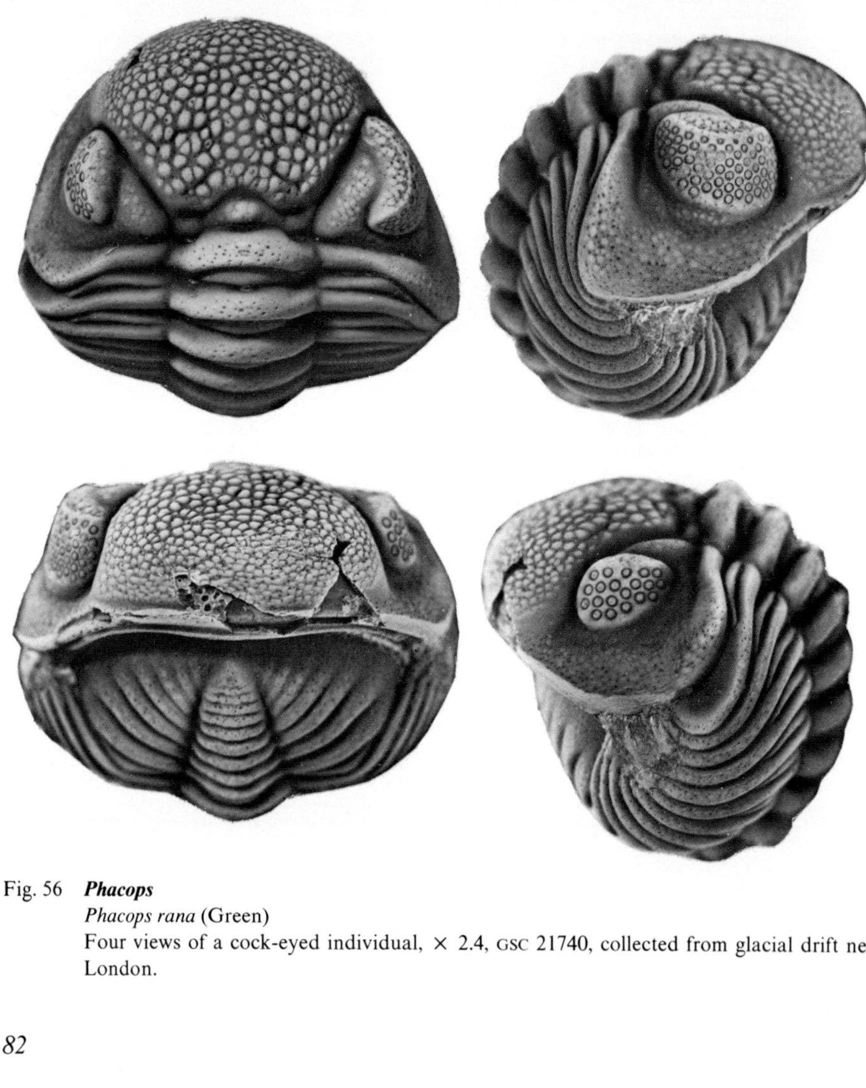

Fig. 56 ***Phacops***
Phacops rana (Green)
Four views of a cock-eyed individual, × 2.4, GSC 21740, collected from glacial drift near London.

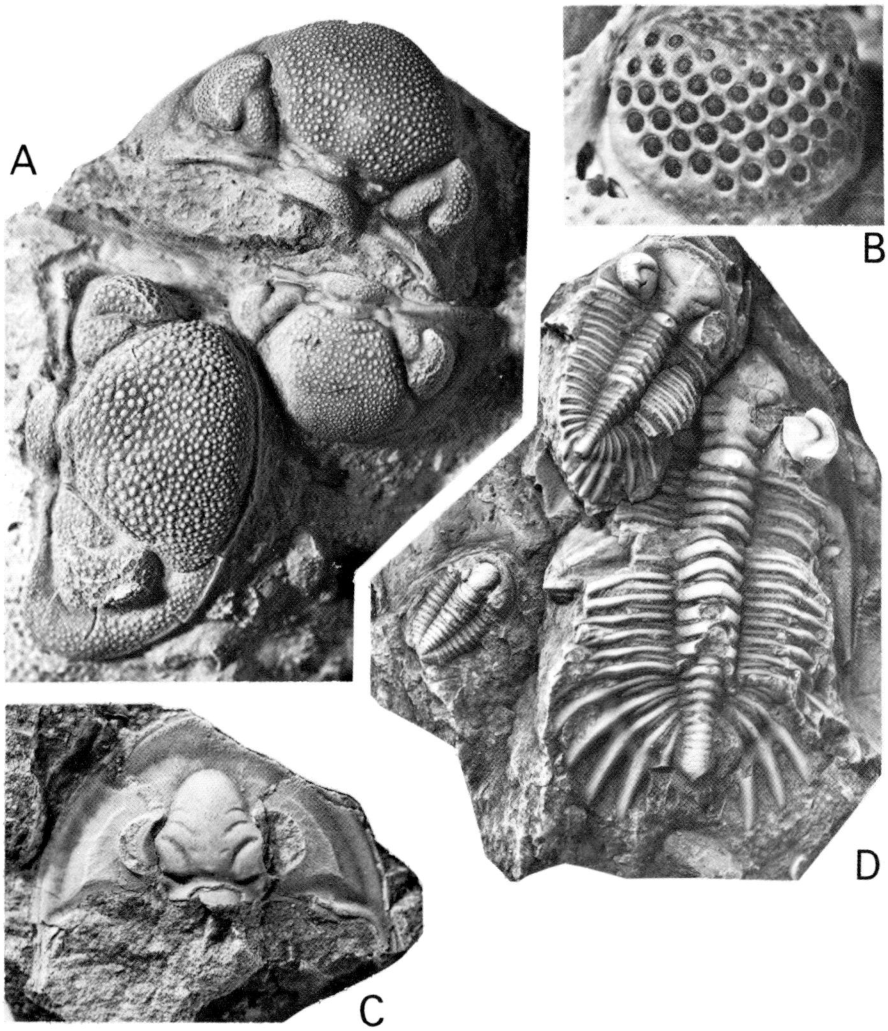

Fig. 57 ***Phacops Dechenella Otarion***

Phacops rana (Green) and *Phacops iowensis* Delo

A Slab with two cephala of *P. rana* (upper right) and one cephalon of *P. iowensis* (lower left), × 1.7, UMMP 24313, Hamilton Gp., Arkona.

B Close-up of eye of *P. rana*, × 5.2, ROM 35400, Hamilton Gp., Moscow, New York.

Dechenella planimarginata (Meek)

C Cephalon, × 1.7, UMMP 15133, Dundee Fm., Trenton, Michigan.

Otarion craspedota (Hall and Clarke) and *Neometacanthus calliteles* (Green)

D Slab with a complete specimen of *O. craspedota* (left) and two specimens of *N. calliteles* (right), × 1.7, NYSM 4235 and 4303, Hamilton Gp., Centrefield, New York.

cas, and Dundee Formations and *Crassiproetus canadensis* Stumm—Hamilton Group, southern Ontario.
REFERENCES Stumm (1953b, 1967)

***Dechenella* Kayser Fig. 57c**
Dechenella has a crescentic and rather flat cephalon with a broad, gently convex border and short pointed genal spines. The pear-shaped glabella has three pairs of curving lateral furrows and there is a tiny lobe located on the lateral portions of the occipital ring. The eyes are large and sausage-shaped. The pygidium is semi-elliptical and has a narrow border and 10 or more distinct ribs.
SPECIES *Dechenella delphinula* (Hall and Clarke)—Bois Blanc Formation, *Dechenella planimarginata* (Meek)—Dundee Formation, *Dechenella alpenensis* Stumm—Hamilton Group, southern Ontario.
REFERENCES Stumm (1953b), Ormiston (1967), Owens (1973)

***Otarion* Zenker Fig. 57D**
This small oval otarionid trilobite has a convex exoskeleton surmounted by a convex axis which is defined by deep furrows. The oval glabella bears a single pair of lateral furrows which isolate a small basal lobe. The preglabellar field is very broad and steeply sloping. *Otarion* has long slender genal spines and a similar median spine on the sixth thoracic segment. The pygidium is tiny and lenticular in outline.
SPECIES *Otarion intermedia* (Weller)—Cataract Group, *Otarion craspedota* (Hall and Clarke)—Hamilton Group, southern Ontario.
REFERENCES Stumm (1967), Whittington and Campbell (1967)

***Dipleura* Green Fig. 58A**
This homalonotid trilobite is very similar to *Trimerus*, differing principally in having a rectangular glabella, facial sutures that diverge markedly behind the eyes, and a less pointed pygidium which lacks obvious segmentation. *Dipleura* is an uncommonly large Devonian trilobite.
SPECIES *Dipleura dekayi* Green—Hamilton Group, southern Ontario.
REFERENCES Cooper (1935), Stumm (1967), Tomczykowa (1975)

***Proetus* Steininger Fig. 58B,C**
Proetus is a moderately inflated proetid trilobite with a relatively broad axis, short genal spines, and a slightly convex border on the cephalon. The glabella lacks lateral furrows. It narrows forwardly and just touches the anterior border. A triangular occipital lobe is located at the juncture of the occipital and axial furrows. The palpebral lobes are large and crescentic and located adjacent to the axial furrows. The semicircular pygidium bears six or more ribs.
SPECIES *Proetus rowi* (Green)—Dundee Formation, *Proetus arkonensis* Stumm —Hamilton Group, southern Ontario.
REFERENCES Hall and Clarke (1888), Stumm (1953b, 1967), Owens (1973)

Fig. 58 ***Dipleura Proetus***
Dipleura dekayi Green
A Exoskeleton, × .9, UMMP 54161, Thunder Bay Limestone, Alpena, Michigan.
Proetus arkonensis Stumm
B Exoskeleton, × 2.2, UMMP 25541, Hamilton Gp., Arkona.
C Exoskeleton, × 2, ROM 1067 H, Hamilton Gp., Arkona.

Repositories of Illustrated Specimens

AMNH	American Museum of Natural History, New York, New York
BM	British Museum (Natural History), London, England
GSC	Geological Survey of Canada, Ottawa
McM	Department of Geology, McMaster University, Hamilton
MCZ	Museum of Comparative Zoology, Harvard University, Cambridge, Massachusetts
NYSM	New York State Museum, Albany, New York
PU	Department of Geology, Princeton University, Princeton, New Jersey
ROM	Royal Ontario Museum, Toronto
UMMP	Museum of Paleontology, University of Michigan, Ann Arbor, Michigan
USNM	United States National Museum of Natural History, Washington, D.C.
UC	Field Museum of Natural History, Chicago, Illinois

Glossary of Morphological Terms

anterior border portion of cephalon in front of preglabellar furrow.

apodeme tongue-shaped extension of outer skeleton on ventral surface; served as muscle attachment area.

axial furrow furrow outlining axial lobe on cephalon, thorax, and pygidium.

axial lobe medial longitudinal region of outer skeleton.

cephalon anterior region of outer skeleton.

coxa interior bladelike portion of appendage located under axial lobe of each segment as well as under cephalon and pygidium.

cranidium central portion of cephalon outlined by facial sutures.

doublure recurved continuation of dorsal outer skeleton onto ventral side as a rim.

exite featherlike gill branch emerging dorsolaterally from coxa.

facial suture bilaterally symmetrical line of weakness along which the cephalon splits during moulting; generally passes along upper edge of eye and in front of preglabellar furrow.

fixed cheek portion of cephalon lying between facial suture and axial furrow.

free cheek lateral portion of cephalon lying outside facial suture.

genal angle posterolateral corner of cephalon.

genal spine posterior extension of posterolateral corner of cephalon.

glabella raised axial portion of cephalon bounded by axial and preglabellar furrows.

holaspid stage final stage in the life history of a trilobite; initiated when the full complement of thoracic segments for that species is attained.

holochroal eye compound eye in which visual surface consists of numerous, densely packed, and small lenses covered by a single continuous cornea.

hypostome ovate or quadrate plate located underneath glabella on ventral side of cephalon.

lateral glabellar furrow furrow extending inward from each side of glabella; occurs in bilaterally symmetrical pairs.

lateral glabellar lobe lateral portion of glabella outlined by successive pairs of lateral glabellar furrows.

marginal suture facial suture with path along margin of cephalon.

meraspid stage second stage in the life history of a trilobite; initiated at the appearance of a joint between the cephalon and the pygidial portion and continues to the holaspid stage.

occipital furrow transverse furrow extending across axial lobe on posterior portion of cephalon.

occipital ring posteriormost portion of axial lobe on cephalon; defined anteriorly by occipital furrow.

opisthoparian suture facial suture with posterior path passing inside genal angle.

palpebral furrow curved furrow that forms inside border of palpebral lobe.

palpebral lobe apronlike extension of fixed cheek that bounds upper edge of eye.

pleural lobe lateral longitudinal region of outer skeleton.

preglabellar furrow furrow defining front edge of glabella.

proparian suture facial suture with posterior path passing in front of genal angle.

protaspid stage earliest stage in the life history of a trilobite; represented by minute undivided oval outer skeleton.

pygidium posterior region of outer skeleton.

rostral plate variably shaped plate located on ventral side of anterior portion of cephalon; generally outlined by sutures on three or four sides.

schizochroal eye compound eye in which visual surface consists of large, separated lenses, each of which is covered by a discrete cornea.

telepod jointed walking leg emerging ventrolaterally from coxa.

thorax middle region of outer skeleton consisting of articulating segments.

trilobite an extinct marine arthropod characterized by a longitudinally trilobed outer skeleton that is divided transversely into anterior, middle, and posterior regions.

Literature Cited

BERGSTRÖM, J.
1973 Organization, life, and systematics of trilobites. Fossils and Strata 2: 1–69.

BEST, R.V.
1961 Intraspecific variation in *Encrinurus ornatus*. Journal of Paleontology 35: 1029–1040.

BIGSBY, J.J.
1824 Notes on the geography and geology of Lake Huron. Transactions of the Geological Society, 2nd ser. 1: 175–208.

BILLINGS, E.
1865 Palaeozoic fossils. Vol. 1. Containing descriptions and figures of new or little known species of organic remains from the Silurian rocks, 1861–1865. Geological Survey of Canada. 426 pp.

BLACK, R.M.
1970 The elements of palaeontology. Cambridge, Cambridge University Press. 339 pp.

BOLTON, T.E.
1957 Silurian stratigraphy and palaeontology of the Niagara Escarpment in Ontario. Geological Survey of Canada, Memoir 289: 1–145.

BOLTON, T.E. and M.J. COPELAND
1972 Paleozoic formations and Silurian biostratigraphy, Lake Timiskaming region, Ontario and Quebec. Geological Survey of Canada, Paper 72-15: 1–49.

BRADLEY, J.H.
1925 Trilobites of the Beekmantown in the Phillipsburg region of Quebec. Canadian Field-Naturalist 39: 5–9.

BRUTON, D.L.
1968 A revision of the Odontopleuridae (Trilobita) from the Palaeozoic of Bohemia. Skrifter Norske Videnskaps-Akademi i Oslo. I. New Zeries 25: 1–73.

CALEY, J.F.
1940 Palaeozoic geology of the Toronto-Hamilton area, Ontario. Geological Survey of Canada, Memoir 224: 1–284.
1941 Palaeozoic geology of the Brantford area, Ontario. Geological Survey of Canada, Memoir 226: 1–176.
1943 Palaeozoic geology of the London area, Ontario. Geological Survey of Canada, Memoir 237: 1–171.

CAMPBELL, K.S.W.

1967 Henryhouse trilobites. Oklahoma Geological Survey, Bulletin 115: 1–68.

1975a The functional morphology of *Cryptolithus*. Fossils and Strata 4: 65–86.

1975b Cladism and phacopid trilobites. Alcheringa 1: 87–96.

1977 Trilobites of the Haragan, Bois d'Arc and Frisco Formations (Early Devonian) Arbuckle Mountains Region, Oklahoma. Oklahoma Geological Survey, Bulletin 123: 1–227.

CHAPMAN, E.J.

1890 Some remarks on the classification of the trilobites, as influenced by stratigraphical relations: with outline of a new grouping of these forms. Transactions of the Royal Society of Canada, sec. 4, 7: 113–120.

CHATTERTON, B.D.E. and R. LUDVIGSEN

1976 Silicified Middle Ordovician trilobites from the South Nahanni River area, District of Mackenzie, Canada. Palaeontographica (Abt. A) 154: 1–106.

CHLUPÁČ, I.

1977 The phacopid trilobites of the Silurian and Devonian of Czechoslovakia. Rozpravy Ústředniho Ústavu Geologického 43: 1–172.

CISNE, J.L.

1973 Life history of an Ordovician trilobite *Triarthrus eatoni*. Ecology 54: 135–142.

1975 Anatomy of *Triarthrus* and the relationships of the Trilobita. Fossils and Strata 4: 45–63.

CLARKE, J.M.

1892 Note on *Coronura aspectans* Conrad (sp.), the *Asaphus diurus*, Green. New York State Geologist, 10th Annual report for 1890, pp. 79–83.

CLARKSON, E.N.K.

1967 Fine structure of the eye in two species of *Phacops* (Trilobita). Palaeontology 10: 603–616.

1975 The evolution of the eye in trilobites. Fossils and Strata 4: 7–31.

CLARKSON, E.N.K. and R. LEVI-SETTI

1975 Trilobite eyes and the optics of Des Cartes and Huygens. Nature 254: 663–667.

COOPER, B.N.

1953 Trilobites from the lower Champlainian formations of the Appalachian Valley. Geological Society of America, Memoir 55: 1–69.

COOPER, G.A.

1935 Young stages of the Devonian trilobite *Dipleura dekayi* Green. Journal of Paleontology 9: 3–5.

CUMMING, L.M.

1975 Ordovician strata of the Hudson Bay Lowlands. Geological Survey of Canada, Paper 74-28: 1–93.

DARBY, D.G. and E.C. STUMM

1965 A revision of the Ordovician trilobite *Asaphus platycephalus* Stokes. Contributions from the Museum of Paleontology, University of Michigan, 20: 63–73.

DELO, D.M.

1940 Phacopid trilobites of North America. Geological Society of America, Special Paper 29: 1–135.

DOUGLAS, R.J.W., ed.
1970 Geology and economic minerals of Canada. Geological Survey of Canada, Economic Geology Report 1: 1–838.

ELDREDGE, N.
1972 Systematics and evolution of *Phacops rana* (Green, 1832) and *Phacops iowensis* Delo, 1935 (Trilobita) from the Middle Devonian of North America. Bulletin of the American Museum of Natural History 147: 47–113.
1973 Systematics of Lower and Lower Middle Devonian species of the trilobite *Phacops* in North America. Bulletin of the American Museum of Natural History 151: 287–337.

EVITT, W.R.
1953 Observations on the trilobite *Ceraurus*. Journal of Paleontology 27: 33–48.

EVITT, W.R. and R.P. TRIPP
1977 Silicified Middle Ordovician trilobites from the families Encrinuridae and Staurocephalidae. Palaeontographica (Abt. A) 157: 129–174.

EVITT, W.R. and H.B. WHITTINGTON
1953 The exoskeleton of *Flexicalymene* (Trilobita). Journal of Paleontology 27: 49–55.

FAGERSTROM, J.A.
1961 The fauna of the Middle Devonian Formosa Reef Limestone of southwestern Ontario. Journal of Paleontology 35: 1–48.

FOERSTE, A.F.
1920 The generic relations of the American Ordovician Lichididae. American Journal of Science 49: 26–50.

FRITZ, M.A.
1964 ?*Scutellum regale* sp. nov. Fritz from the Silurian of the Hudson Bay area. Proceedings of the Geological Association of Canada, 15, pt. 2: 91–97.

HALL, J. and J.M. CLARKE
1888 Trilobites and other crustacea of the Oriskany, upper Helderberg, Hamilton, Portage, Chemung and Catskill Groups. Palaeontology of New York 7. 236 pp.

HEWITT, D.F.
1972 Paleozoic geology of southern Ontario. Ontario Division of Mines, Geological Report 105: 1–18.

HUGHES, C.P., J.K. INGHAM and R. ADDISON
1975 The morphology, classification and evolution of the Trinucleidae (Trilobita). Philosophical Transactions of the Royal Society of London, pt. B, 272: 537–607.

JAANUSSON, V.
1954 Zur Morphologie und Taxonomie der Illaeniden. Arkiv för Mineralogi och Geologi 1: 545–583.

KESLING, R.V. and R.B. CHILMAN
1975 Strata and megafossils of the Middle Devonian Silica Formation. University of Michigan, Museum of Paleontology, Papers on Paleontology, 8: 1–408.

LANE, P.D.
1971 British Cheiruridae (Trilobita). Palaeontographical Society, Monograph 125: 1–95.

LESPÉRANCE, P.J.
1975 Stratigraphy and paleontology of the Synphoriidae (Lower and Middle Devonian dalmanitacean trilobites). Journal of Paleontology 49: 91–137.

LESPÉRANCE, P.J. and P.A. BOURQUE

1971 The Synphoriinae: an evolutionary pattern of Lower and Middle Devonian trilobites. Journal of Paleontology 45: 182–208.

LEVI-SETTI, R.

1975 Trilobites: a photographic atlas. Chicago, University of Chicago Press. 213 pp.

1976 Ancient and wonderful eyes. Fossils Magazine 1(1): 24–34.

LIBERTY, B.A.

1957 Manitoulin Island, District of Manitoulin, Ontario. Geological Survey of Canada, Map 1957-20.

1964 Middle Ordovician stratigraphy of the Lake Simcoe area, Ontario. *In* Copeland, M.J., ed. Guidebook, geology of central Ontario, Toronto. American Association of Petroleum Geologists and Society of Economic Paleontologists and Mineralogists, pp. 15–24.

1969 Palaeozoic geology of the Lake Simcoe area, Ontario. Geological Survey of Canada, Memoir 355: 1–201.

LIBERTY, B.A. and T.E. BOLTON

1971 Paleozoic geology of the Bruce Peninsula area, Ontario. Geological Survey of Canada, Memoir 360: 1–163.

LIBERTY, B.A. and F.D. SHELDEN, eds.

1968 The geology of Manitoulin Island. Michigan Basin Geological Society Annual Field Trip Excursion Guidebook. 101 pp.

LOGAN, W.E.

1863 Geology of Canada: report of progress from its commencement to 1863. Montreal, Geological Survey of Canada. 983 pp.

LUDVIGSEN, R.

1977 The Ordovician trilobite *Ceraurinus* Barton in North America. Journal of Paleontology 51: 959–972.

1978 The trilobites *Bathyurus* and *Eomonorachus* from the Middle Ordovician of Oklahoma and their biofacies significance. Royal Ontario Museum, Life Sciences Contribution 114: 1–18.

In press The Ordovician trilobite *Pseudogygites* Kobayashi in eastern and arctic North America. Royal Ontario Museum, Life Sciences Contribution.

MacFALL, R.P. and J.C. WOLLIN

1972 Fossils for amateurs: a guide to collecting and preparing invertebrate fossils. New York, Van Nostrand Reinhold. 341 pp.

MILLER, J.

1976 The sensory fields and life mode of *Phacops rana* (Green, 1832) (Trilobita). Transactions of the Royal Society of Edinburgh 69: 337–367.

MOORE, R.C., ed.

1959 Treatise on invertebrate paleontology; part O, Arthropoda 1. Lawrence, Geological Society of America and University of Kansas Press. 560 pp.

MOORE, R.C., C.G. LALICKER and A.G. FISCHER

1952 Invertebrate fossils. New York, McGraw-Hill. 766 pp.

NICHOLSON, H.A.

1874 Report upon the palaeontology of the province of Ontario. Toronto, Hunter Rose. 133 pp.

1875 Report upon the palaeontology of the province of Ontario. Toronto, Hunter Rose. 96 pp.

ORMISTON, A.R.
1967 Lower and Middle Devonian trilobites of the Canadian Arctic Islands. Geological Survey of Canada, Bulletin 153: 1–148.

OWENS, R.M.
1973 British Ordovician and Silurian Proetidae (Trilobita). Palaeontographical Society, Monograph 127: 1–98.

PARKS, W.A.
1928 Faunas and stratigraphy of the Ordovician black shales and related rocks in southern Ontario. Transactions of the Royal Society of Canada, sec. 4, 22: 39–92.

PERRY, D.G. and B.D.E. CHATTERTON
1977 Silurian (Wenlockian) trilobites from Baillie-Hamilton Island, Canadian Arctic Archipelago. Canadian Journal of Earth Sciences 14: 285–317.

PHLEGER, F.B.
1937 New Lichidacea in the collections of the Museum of Comparative Zoology. Bulletin of the Museum of Comparative Zoology 80: 415–423.

PŘIBYL, A. and J. VANĚK
1971 Studie über die Familie Scutelluidae Richter et Richter (Trilobita) und ihre phylogenetische Entwicklung. Acta Universitatis Carolinae—Geologica no. 4: 361–394.

RAYMOND, P.E.
1916 New and old Silurian trilobites from southeastern Wisconsin, with notes on the genera of the Illaenidae. Bulletin of the Museum of Comparative Zoology 60: 1–41.
1921 A contribution to the description of the fauna of the Trenton Group. Geological Survey of Canada, Museum Bulletin 31: 1–68.
1924 New Upper Cambrian and Lower Ordovician trilobites from Vermont. Proceedings of the Boston Society of Natural History 37: 389–466.

RAYMOND, P.E. and D.C. BARTON
1913 A revision of the American species of *Ceraurus*. Bulletin of the Museum of Comparative Zoology 54: 525–543.

RAYMOND, P.E. and J.E. NARRAWAY
1906 Notes on Ordovician trilobites: Illaenidae from the Black River Limestone near Ottawa, Canada. Annals of the Carnegie Museum 4: 242–255.

REIMANN, I.G.
1945 The thorax of *Terataspis*. Journal of Paleontology 19: 69–71.

ROSS, R.J., JR.
1967 Calymenid and other Ordovician trilobites from Kentucky and Ohio. United States Geological Survey, Professional Paper 583-B: 1–18.

SANFORD, B.V.
1968 Devonian of Ontario and Michigan. *In* Oswald, D.H., ed. International symposium on the Devonian System. Calgary, Alberta Society of Petroleum Geologists, pp. 973–999.

SANFORD, B.V. and W.B. BRADY
1955 Palaeozoic geology, Windsor-Sarnia area, Ontario. Geological Survey of Canada, Memoir 278: 1–65.

SANFORD, B.V. and A.W. NORRIS

1973 The Hudson Platform. *In* McCrossan, R.G., ed. The future petroleum provinces of Canada: their geology and potential. Canadian Society of Petroleum Geologists, Memoir 1: 387–409.

1975 Devonian stratigraphy of the Hudson Platform. Geological Survey of Canada, Memoir 379: I, 1–124; II, 1–248.

SEGALL, R.T., ed.

1972 Niagaran stratigraphy: Hamilton, Ontario. Michigan Basin Geological Society Annual Field Trip Excursion Guidebook. 89 pp.

SHAW, F.C.

1968 Early Middle Ordovician Chazy trilobites of New York. New York State Museum and Science Service, Memoir 17: 1–163.

1974 Simpson Group (Middle Ordovician) trilobites of Oklahoma. Paleontological Society, Memoir 6: 1–54.

SHIRLEY, J.

1936 Some British trilobites of the family Calymenidae. Quarterly Journal of the Geological Society of London 92: 384–422.

SINCLAIR, G.W.

1944 Some Ordovician trilobites from Ontario. Transactions of the Royal Canadian Institute 25: 15–20.

1946 The Ordovician trilobite *Dimeropyge*. American Journal of Science 244: 854–860.

1947 Muscle scars in the Ordovician trilobite *Illaenus*. American Journal of Science 245: 529–536.

1949 The Ordovician trilobite *Eobronteus*. Journal of Paleontology 23: 45–56.

SIVETER, D.J.

1976 The Middle Ordovician of the Oslo region, Norway, 27. Trilobites of the family Calymenidae. Norsk Geologisk Tidsskrift 56: 335–396.

ŠNAJDR, M.

1960 A study of the family Scutelluidae (Trilobitae). Rozpravy Ústředního Ústavu Geologického 26: 1–264.

STAUFFER, C.R.

1915 The Devonian of southwestern Ontario. Geological Survey of Canada, Memoir 34: 1–341.

STUBBLEFIELD, C.J.

1926 Notes on the development of a trilobite, *Shumardia pusilla* (Sars). Journal of the Linnean Society, Zoology 36: 345–372.

STUMM, E.C.

1953a Trilobites of the Devonian Traverse Group of Michigan. Contributions from the Museum of Paleontology, University of Michigan, 10: 101–157.

1953b Lower Middle Devonian proetid trilobites from Michigan, southwestern Ontario, and northern Ohio. Contributions from the Museum of Paleontology, University of Michigan, 11: 11–31.

1954 Lower Middle Devonian phacopid trilobites from Michigan, southwestern Ontario, and the Ohio Valley. Contributions from the Museum of Paleontology, University of Michigan, 11: 201–221.

1967 Devonian trilobites from northwestern Ohio, northern Michigan, and western New York. Contributions from the Museum of Paleontology, University of Michigan, 21: 109–122.

STUMM, E.C. and E.G. KAUFFMAN

1958 Calymenid trilobites from the Ordovician rocks of Michigan. Journal of Paleontology 32: 943–960.

STUMM, E.C. and J.D. WRIGHT

1958 Check list of fossil invertebrates described from the Middle Devonian rocks of the Thedford-Arkona region of southwestern Ontario. Contributions from the Museum of Paleontology, University of Michigan, 14: 81–132.

STÜRMER, W. and J. BERGSTRÖM

1973 New discoveries on trilobites by X-rays. Paläontologische Zeitschrift 47: 104–141.

THOMAS, A.T.

1977 Classification and phylogeny of homalonotid trilobites. Palaeontology 20: 159–178.

TOMCZYKOWA, E.

1975 The trilobite subfamily Homalonotinae from the Upper Silurian and Lower Devonian of Poland. Acta Palaeontologica Polonica 20: 3–46.

TOWE, K.M.

1973 Trilobite eyes: calcified lenses in vivo. Science 179: 1007–1009.

TRIPP, R.P.

1957 The classification and evolution of the superfamily Lichacea (Trilobita). Geological Magazine 94: 104–122.

1962 The Silurian trilobite *Encrinurus punctatus* (Wahlenberg) and allied species. Palaeontology 5: 460–477.

TRIPP, R.P., J.T. TEMPLE and K.C. GASS

1977 The Silurian trilobite *Encrinurus variolaris* and allied species, with notes on *Frammia*. Palaeontology 20: 847–867.

WARBURG, E.

1939 The Swedish Ordovician and Lower Silurian Lichidae. Kungliga Svenska Vetenskapsakademiens Handlingar, 3rd ser., 17 (4): 1–162.

WELLER, S.

1909 The paleontology of the Niagaran Limestone in the Chicago area. The Trilobita. Chicago Academy of Sciences, Natural History Survey Bulletin 4, pt. 2: 163–281.

WHITTINGTON, H.B.

1941 Silicified Trenton trilobites. Journal of Paleontology 15: 492–522.

1949 *Dolichoharpes* and the origin of the harpid fringe. American Journal of Science 247: 276–285.

1950 A monograph of the British trilobites of the family Harpidae. Palaeontographical Society, Monograph 103: 1–55.

1952 A unique remopleuridid trilobite. Breviora 4: 1–9.

1953 North American Bathyuridae and Leiostegiidae (Trilobita). Journal of Paleontology 27: 647–678.

1954 Ordovician trilobites from Silliman's Fossil Mount. *In* Miller, A.K. *et al.* Ordovician Cephalopods of Baffin Island. Geological Society of America, Memoir 62: 119–149.

1956 Silicified Middle Ordovician trilobites: the Odontopleuridae. Bulletin of the Museum of Comparative Zoology 114: 155–288.

1957 Ontogeny of *Elliptocephala, Paradoxites, Sao, Blainia*, and *Triarthrus*. Journal of Paleontology 31: 934–946.

1959 Silicified Middle Ordovician trilobites: Remopleurididae, Trinucleidae, Raphiophoridae, Endymionidae. Bulletin of the Museum of Comparative Zoology 121: 371–496.

1960 *Cordania* and other trilobites from the Lower and Middle Devonian. Journal of Paleontology 34: 405–420.

1968 *Cryptolithus* (Trilobita): specific characters and occurrence in Ordovician of eastern North America. Journal of Paleontology 42: 702–714.

1971 Silurian calymenid trilobites from United States, Norway, and Sweden. Palaeontology 14: 455–577.

1975 Trilobites with appendages from the Middle Cambrian, Burgess Shale, British Columbia. Fossils and Strata 4: 97–136.

WHITTINGTON, H.B. and K.S.W. CAMPBELL

1967 Silicified Silurian trilobites from Maine. Bulletin of the Museum of Comparative Zoology 135: 447–482.

WHITTINGTON, H.B. and W.R. EVITT

1954 Silicified Middle Ordovician trilobites. Geological Society of America, Memoir 59: 1–137.

WILSON, A.E.

1946 Geology of the Ottawa-St. Lawrence Lowland, Ontario and Quebec. Geological Survey of Canada, Memoir 241: 1–65.

1947 Trilobita of the Ottawa Formation of the Ottawa-St. Lawrence Lowland. Geological Survey of Canada, Bulletin 9: 1–86.

WILSON, A.E., J.F. CALEY, J.C. SPROULE and V.J. OKULITCH

1936 Contributions to the study of the Ordovician of Ontario and Quebec. Geological Survey of Canada, Memoir 202: 1–133.

WINDER, C.G., C.R. BARNES, P.G. TELFORD and T.T. UYENO

1975 Ordovician to Devonian stratigraphy and conodont biostratigraphy of southern Ontario [Guidebook]. Geological Association of Canada Annual Meeting, Waterloo, pp. 119–160.

WINDER, C.G. and B.V. SANFORD

1972 Stratigraphy and paleontology of the Paleozoic rocks of southern Ontario. Guidebook to Excursion A45-C45, 24th International Geological Congress, Montreal. 74 pp.